Springer Theses

Recognizing Outstanding Ph.D. Research

Aims and Scope

The series "Springer Theses" brings together a selection of the very best Ph.D. theses from around the world and across the physical sciences. Nominated and endorsed by two recognized specialists, each published volume has been selected for its scientific excellence and the high impact of its contents for the pertinent field of research. For greater accessibility to non-specialists, the published versions include an extended introduction, as well as a foreword by the student's supervisor explaining the special relevance of the work for the field. As a whole, the series will provide a valuable resource both for newcomers to the research fields described, and for other scientists seeking detailed background information on special questions. Finally, it provides an accredited documentation of the valuable contributions made by today's younger generation of scientists.

Theses may be nominated for publication in this series by heads of department at internationally leading universities or institutes and should fulfill all of the following criteria

- They must be written in good English.
- The topic should fall within the confines of Chemistry, Physics, Earth Sciences, Engineering and related interdisciplinary fields such as Materials, Nanoscience, Chemical Engineering, Complex Systems and Biophysics.
- The work reported in the thesis must represent a significant scientific advance.
- If the thesis includes previously published material, permission to reproduce this must be gained from the respective copyright holder (a maximum 30% of the thesis should be a verbatim reproduction from the author's previous publications).
- They must have been examined and passed during the 12 months prior to nomination.
- Each thesis should include a foreword by the supervisor outlining the significance of its content.
- The theses should have a clearly defined structure including an introduction accessible to new PhD students and scientists not expert in the relevant field.

Indexed by zbMATH.

More information about this series at https://link.springer.com/bookseries/8790

James A. Gott

Defects in Self-Catalysed III-V Nanowires

Doctoral Thesis accepted by
University of Warwick, Coventry, United Kingdom

Springer

Author
Dr. James A. Gott
Department of Physics
University of Warwick
Coventry, UK

Supervisor
Prof. Ana M. Sanchez
Department of Physics
University of Warwick
Coventry, UK

ISSN 2190-5053 ISSN 2190-5061 (electronic)
Springer Theses
ISBN 978-3-030-94061-4 ISBN 978-3-030-94062-1 (eBook)
https://doi.org/10.1007/978-3-030-94062-1

This Springer imprint is published by the registered company Springer Nature Switzerland AG
The registered company address is: Gewerbestrasse 11, 6330 Cham, Switzerland

Supervisor's Foreword

Semiconductor materials are well-known for having a key role in the modern world. The continuous demand to reduce the footprint of devices has inspired the development of novel optoelectronic architectures. In principle, one-dimensional nanowires have an inherent ability to grow as perfect crystals, thanks to the strong surface image forces that act to expel any defect with a long-range strain field. There is no need for a nanowire crystal structure to match that of the substrate, eliminating the severe lattice parameter-matching constraints that so limit the range of epitaxial layers on bulk substrates. Thus, semiconductor nanowires promise the ease of integration of III-V materials with well-established silicon technology.

James Gott's Ph.D. thesis presents methodical research using electron microscopy to describe previously undiscovered, topologically protected, line defects with the null Burgers vector that, unlike dislocations, are stable in nanoscale crystals. These are studied at atomic resolution using aberration corrected electron microscopy. A Burgers circuit protocol is developed that can be applied to multiply twinned material without the problems or ambiguity that affect other approaches. The defect microstructure is quite different from that seen in bulk material, but they also quench the luminescence. In-situ electron microscopy at the atomic level using short annealing cycles was employed to determine the stability and dynamics of these defects. The majority are unstable when exposed to high temperature, suggesting that in-situ annealing would considerably improve nanowire quality. These results are extremely important for the nanowire community since they provide strategies to enhance the structural properties of self-catalysed nanowires after droplet consumption. This Ph.D. thesis also examined the interfaces in axial quantum dots in semiconducting nanowires. Different theoretical and experimental models were considered in the analysis.

The work by James Gott represents a breakthrough in discovering and investigating topologically protected line defects in nanowires, a system which was thought to be intrinsically defect-free. He extracted the characteristic signatures of these defects using beautifully resolved atomic images. His Ph.D. thesis is a clear example of how electron microscopy will continue to fascinate physicists and materials scientists for years to come.

Warwick, UK
August 2021

Prof. Ana M. Sanchez

Abstract

Semiconductor nanowires are poised to be a candidate for next-generation technology with superior performance and high integration ability. They have unique physical properties that are enabled by their nanoscale form-factor. Nanowires are commonly described as being defect-free due to their ability to expel mobile defects with long-range strain fields. The droplet consumption step in self-catalysed III-V nanowires can produce material with a high density of line defects, often with the null Burgers vector, i.e., no long-range strain field. The presence of such defects can diminish device performance and make them unreliable.

This thesis presents an extensive study made into defects present in semiconductor nanowires. Defect structures are analysed from atomic resolution electron microscope images, and observations show that the nanowire microstructure is very different from bulk material. Nanowires can contain line defects that (a) are trapped by locks or other defects, (b) arranged as dipoles or groups with a zero total Burgers vector, or (c) have a zero Burgers vector. The most common defect is the three-monolayer high twin facet with a zero Burgers vector. Cathodoluminescence experiments reveal optical emission is quenched in defective regions, showing that they act as strong non-radiative recombination centers.

The stability of defects is tested by in-situ electron microscopy to analyse defect behaviour in GaAsP nanowires using short annealing cycles. Movement of the null Burgers vector defects appears to be consistent with the thermally activated single- or double-kink mechanisms of dislocation glide, with velocities that do not exceed 1 nm s^{-1}. The motion of the null Burgers vector defects is found to depend on their size, position, and surrounding environment and sets an upper limit to activation energy around 2 eV. The majority of defects (>70%) are removed by post-growth annealing for several seconds at temperatures in excess of 640 °C, while the remaining defects do not move and are thermodynamically stable in the nanowire.

Finally, axial heterostructures in GaAsP nanowires with GaAs quantum dots are examined and a selection of theoretical models are tested to see how well they fit experimental interfaces. Of the physical models tested, a model recently developed by Dubrovskii was found to best fit experimental data. Interface sharpness and size of quantum dots were measured for different nanowire radii and both showed no obvious trend. These results are explained by the low group V concentration and solubility in the catalyst, and length distribution of the nanowires, respectively.

Acknowledgements

Along my Ph.D. journey, there have been a number of people who have been instrumental in my progression and whose help has made this Ph.D. thesis possible.

Firstly, I am incredibly grateful to my supervisor, Prof. Ana Sanchez, who has countless times provided invaluable guidance professionally, academically and also personally. Her microscopy reputation precedes her and it has been a privilege to learn from one of the best. Her dedication to her work and the people she works with is admirable and inspiring.

Prof. Richard Beanland has been an influential figure in my career and without whom I would not be where I am today. His wealth of knowledge in microscopy and crystal defects is unrivalled, and I thank him for his help and guidance over the years. Prof. Neil Wilson and Dr. Jeremy Sloan must be thanked for their contributions to the group and for making the group complete.

I thank Dr. Jon Peters for being a (mostly) great role model and a good friend. He has provided an enormous amount of help over the years, from practical microscopy to helping with programming. He has made a significant contribution through my Ph.D. time and has played some part in moulding the person I am today.

I must thank Dr. Aruni Fonseka for all the fruitful discussions, guidance and help she has provided. Her knowledge and experience in nanowires and nanowire growth have proven to be incredibly useful, and I have learnt a great deal from her academic experience. She should also be thanked for teaching me how to use the ultramicrotome.

I have had the pleasure of sharing my Ph.D. journey with Charlie Slade, a colleague and friend I dearly appreciate. I thank her for her support when my worrying gets the better of me.

I would like to acknowledge colleagues and friends, past and present, who have made the Ph.D. journey a more enjoyable experience: Sam Marks, Zachary Laker, Natalie Teutsche, Bo Li, Xue Xia, Fraser Laidlaw, Abi Graham and many others.

Mention should be made to project collaborators, especially Dr. Yunyan Zhang (UCL) who has always been helpful and a great source of knowledge on nanowires, and without him, there would be no nanowires to study in this thesis. Prof. Huiyun Liu (UCL) and Prof. David Mowbray (Sheffield) have provided valuable knowledge and insight into both their respective fields and the world of academia. Both have proven to be wonderful collaborators.

Praise and appreciation must be given to the microscopy RTP staff, Steve York and Steve Hindmarsh, who have always been helpful and do an excellent job at keeping the microscopes humming along. Particular thanks go to Steve Hindmarsh for all the help with the cryo-CL system.

Finally, I would like to thank my family, and particularly my parents, for being supportive and being a source of encouragement for the duration of my Ph.D.

Contents

Acronyms

ADF	Annular dark-field
ARM	Atomic Resolution analytical Microscope
BF	Bright-field
CBED	Convergent beam electron diffraction
CCD	Charge-coupled device
CEOS	Corrected electron optical systems
CL	Cathodoluminescence
CSL	Coincidence site lattice
CVD	Chemical vapour deposition
DFT	Density functional theory
DP	Diffraction pattern
EDX	Energy-dispersive X-ray spectroscopy
E-T	Everhart-Thornley
FCC	Face-centred cubic
FET	Field effect transistor
FFT	Fast Fourier transform
GPA	Geometric phase analysis
HAADF	High angle annular dark field
HCP	Hexagonal close packed
IFFT	Inverse fast Fourier transform
LCL	Lomer-Cottrell lock
LED	Light emitting diode
MBE	Molecular beam epitaxy
ML	Monolayer
NW	Nanowire
PMT	Photo multiplier tube
QD	Quantum dot
QW	Quantum well
RHEED	Reflection high-energy electron diffraction
RTD	Resonant tunnel diode
SC	Semiconductor

SE	Secondary electron
SEM	Scanning electron microscope
SIMS	Secondary ion mass spectrometry
STEM	Scanning transmission electron microscope
STM	Scanning tunneling microscopy
TDS	Thermal diffuse scattering
TEM	Transmission electron microscope
VLS	Vapour-liquid-solid
VS	Vapour-solid
WZ	Wurtzite
ZB	Zincblende

Chapter 1
Introduction

1.1 Introduction

Semiconductor (SC) materials and devices play a fundamental role in the modern world. Despite their success and prominence, one issue that can still cause problems is the inability to perfectly combine mismatched material systems. Difficulty in combining different materials comes from lattice mismatch and differences in thermal expansion coefficients. These differences can generate dislocations that prevent devices from operating or make them unreliable.

SC nanowires (NWs) represent a significant advance toward seamless integration of highly mismatched materials with silicon. There is interest in NWs because of their inherent ability, in principle, to grow as perfect crystals. The crystal perfection in NWs is a result of their small volume and relatively large surface area which produces strong image forces on any defect with a long range strain field. The ability of NWs to expel line defects overcomes key challenges in device production. The inherent perfection of NWs may allow production of SC materials and devices without dislocations or cracks, which are currently unobtainable using existing heteroepitaxial layer growth techniques.

While NW growth procedures are improving, one of the remaining difficulties is how to remove the catalyst droplet once growth is finished. In self-catalysed GaAs NWs, the Ga droplet can be consumed to form solid GaAs, via a different growth mode. Other growth modes constitute sub-optimal growth conditions that lead to defective structures, including line defects with no strain that can act as non-radiative recombination centres [1].

This thesis employs electron microscopy, especially the transmission electron microscope (TEM), to explore structural defects found in self-catalysed III–V NWs primarily through atomic resolution imaging. Type of defect structures are identified by analysing Burgers vectors, and defect dynamics are investigated using in-situ experiments. Axial heterostructures in GaAsP NWs with GaAs quantum dots (QDs) are also investigated, where models that describe heterostructure composition

J. A. Gott, *Defects in Self-Catalysed III-V Nanowires*, Springer Theses,
https://doi.org/10.1007/978-3-030-94062-1_1

changes are considered and some characteristics are tested to see if they vary with NW radius as expected.

1.2 Nanowires

1.2.1 Introduction to Nanowires

NWs have become a topic of increasingly intense study over the last 20 years [2–12]. The term covers a wide range of materials that allow for a variety of nanostructures to be fabricated. As a rough definition, they cover materials with a diameter of up to a few hundred nm, and length of up to a few μm, as illustrated in Fig. 1.1a. Challenges that have followed this field include how to control shape, size and composition of nanowires. Advances in the field of NWs have proven that many of these parameters can be controlled to a high degree of accuracy. Motivation behind advances are largely around their potential applications in devices and chasing further miniaturisation of electronics [12, 13], photonics [11, 14], energy conversion and storage [15].

As an example of NWs, an experimental secondary electron (SE) scanning electron microscope (SEM) image of some GaAsP NWs attached to the Si (111) substrate they were grown on are shown in Fig. 1.1b. This image demonstrates typical features of growth, where unintentional islands of epi-layer growth accompany NWs on the substrate.

Foundations for research into the field of NWs includes growing thin film or 2D layered structures using molecular beam epitaxy (MBE) and also QD structures grown on large substrates [16]. Early examples of NW fabrication include the synthesis of Si wires in 1957 [17] and Si nanoribbons [18]. The commonly used

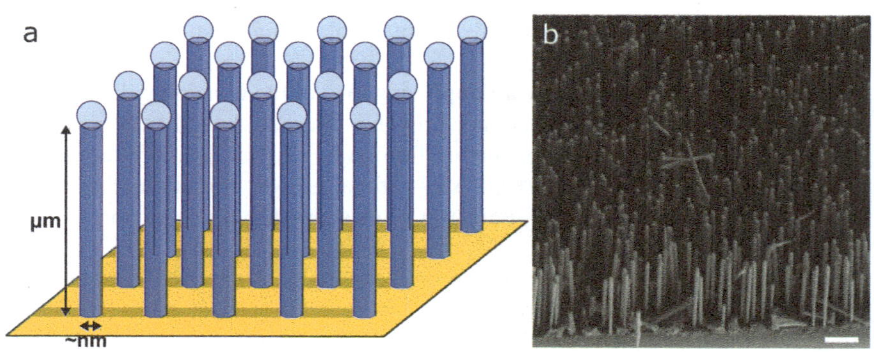

Fig. 1.1 **a** Illustration of NWs attached to a substrate with catalyst droplets attached. NWs are typically up to a few hundred nms wide and up to a few microns tall. **b** SE SEM image of NWs attached to substrate. Accelerating voltage 3 kV. Scale bar 1 μm

vapour-liquid-solid (VLS) growth method was first established by Wagner and Ellis shortly after [19]. It was soon realised that NW geometry and the VLS mechanism allow for the combination of materials different to that of thin film or bulk materials [20]. Work performed by the likes of Lieber et al. [21] and Wu et al. [22] demonstrated the potential of the VLS regime to grow high quality NW structures. The growth of NWs is a very complicated and intricate topic, and the fine details of the physics behind growth is still being studied today [23–25].

1.2.2 Nanowire Growth

A range of techniques have been used to grow NW structures, including: laser ablation [26], metalorganic vapour-phase epitaxy [27], chemical beam epitaxy [28], MBE [29], supercritical fluid-liquid-solid [30], solution-liquid-solid [31, 32], vapour-solid-solid [33] and oxide assisted growth [34] to name a few. Growth broadly utilises one of two approaches, the bottom-up or the top-down [35, 36]. The bottom-up approach involves structure assembly from their components in an additive fashion [35]. The top-down approach uses sculpting or etching to carve structures from a larger piece of material in a subtractive fashion. The top-down approach is used more in integrated devices [35], with more defective surfaces and poor control over facets [12].

The bottom-up approach typically allows for better handling of strain [11] and has a reputation for superior material quality [12]. There are two main varieties of bottom-up growth; catalyst-free and catalytic growth. Catalyst-free growth uses the vapour-solid (VS) mechanism and usually involves masking the substrate [11]. Catalytic growth involves using a nanoparticle catalyst to assist NW growth, and can be further categorised into foreign metal catalysed and self-catalysed growth [37]. The bottom-up approach can also use templates to control position and size of NWs on the substrate, with this technique used in III–V integration on Si [38–41].

In the catalyst assisted growth mode, the catalyst droplet acts as both a source material collector and as the reaction catalyst [11, 42]. During growth, the droplet collects material at the vapour-liquid interface [2], which comes from direct impingement of vapour material from a source, and diffusion of material along NW sidewalls and substrate [23]. A schematic illustration of catalyst assisted growth is shown in Fig. 1.2. When the catalyst droplet is oversaturated, NW growth proceeds by material precipitation at the liquid-solid interface [2]. The most commonly used foreign metal catalyst is gold [11, 19, 37, 42, 43]. Au catalysed III–V growth has shown success in high quality structures grown, however a disadvantage is the gold can be incorporated into the grown structure, contaminate the growth, and alter physical properties.

Alternatively, as was predicted early on [19], it is possible to use a component of the material being grown as the catalyst [44, 45]. Self-catalysed growth has advantages of being able to produce high purity growth and gives the ability to control size of NW by controlling catalyst size. It is also possible to 'consume' the catalyst droplet towards the end of growth by adjusting growth conditions [11]. This crys-

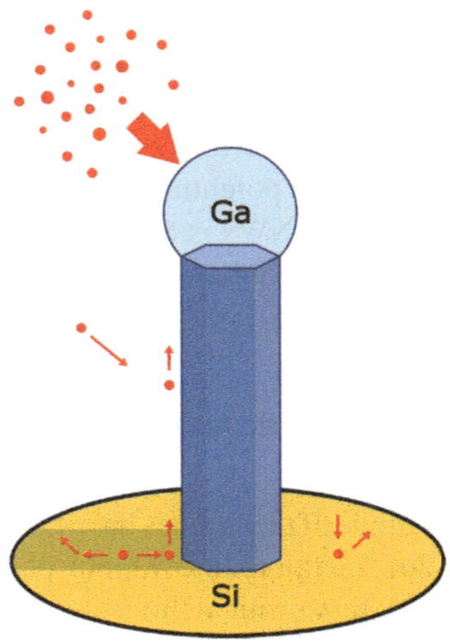

Fig. 1.2 Schematic of VLS NW growth. The example shown here is a self-catalysed GaAs NW on a Si substrate

tallises the droplet which then prevents further axial growth, and the option of only radial growth via the VS mechanism becomes available.

The nature of NW growth has been probed by the use of in-situ electron microscopy, which has allowed for the observation of NWs during growth inside a TEM [46–48]. The technique provides insight into nucleation, catalyst stability, surface structure, and growth kinetics.

In typical catalyst-assisted growth, the substrate is first prepared by distributing catalyst droplets that dictate position and diameter of NWs grown [12]. It has been shown in self-catalysed GaAs NWs that the length of NWs follows a subpoissonian distribution [49, 50].

The growth process is dictated by a few key parameters. These include the radius of NW, the catalyst contact angle at the solid-liquid interface, temperature, and the thermodynamic properties of the material systems relevant to each particular growth [44, 45, 51, 52]. In III–V systems, because of low surface energy, the $\langle 111 \rangle$ growth direction is favoured. In most cases for polar NWs, $\langle 111 \rangle$B polarity (group V terminated) is preferred in both III–V and II–VI materials because of a small surface energy [53]. The polarity of the material can influence NW properties, including structural defect formation in growth, impurity incorporation, and electronic properties of the NWs [6]. $\langle 111 \rangle$B NWs tend to be more tapered and defective while the opposite is true for $\langle 111 \rangle$A (group III terminated) NWs. It has been shown that growth direction and polarity can be controlled by adjusting reactant concentrations and catalyst contact angle [54, 55].

III–V NWs can be grown with either zincblende (ZB) or wurtzite (WZ) crystal structure. WZ NW sidewalls have a lower surface energy. The energy of a surface is related to its surface free energy, which is the work that can be obtained from destruction of a unit area of the surface [56]. NWs are usually bounded by $\{110\}$ and $\{112\}$ side facets due to low surface energy [57]. A feature of the NW system is the ability to grow polytype structures. Hiruma et al. [58] found that it is possible to change growth of GaAs 'whiskers' from ZB to WZ by adjusting growth conditions, which means it is possible to control crystal phase in NWs. This is in contrast to bulk materials, where for most materials there only exists one stable phase. Efforts were made in developing theoretical understanding of why crystal phase can change during growth, along with which materials and what conditions are needed to achieve phase switching [59, 60]. The control of phase switching was demonstrated by Dick et al. [61], and precise control was achieved by Lehmann et al. [62]. Experimental evidence for how phase relies on various parameters was found by Jacobsson et al. [47] where in-situ experiments were performed. GaAs NWs were imaged during growth using Au catalyst inside a TEM, and it was shown that droplet geometry is a key parameter because of its effect on NW edge morphology. During growth, two phases can be switched between by changing the ratio of precursor gases, with WZ GaAs seen to form with higher V/III ratio. A similar experiment has recently been performed by Panciera et al. [63] in self-catalysed GaAs NWs and similar behaviour was reported.

It has been shown there is a difference in the growth dynamics between ZB and WZ phases, with WZ growing by step flow across the interface. This type of step flow was seen early in in-situ NW growth studies on Si NWs [64]. The steps flow slowly with each step starting as soon as the previous one has completed. Recent in-situ TEM experiments have shown that this occurs in WZ phase growth of Au catalysed GaAs NW growth. Harmand et al. showed steps nucleate at the edge of the NW and progress with step flow over the course of a few seconds [48]. ZB growth also proceeds one layer at a time however it is much faster [47].

NWs have potential for more complicated growth. It has been shown that the high surface to volume ratio of NWs can be pushed further by second nucleation on the surface of NWs [13]. Subsequent growth on NW walls produces branched structures, and this was first demonstrated by Dick et al. [65].

1.2.3 Physical Properties

NWs have an established reputation for superior performance in a variety of applications because of more desirable physical properties. This section looks to explore some of the physical properties of NWs and covers those that make them a desirable structure to fabricate.

One desirable property is the ability to confine charge carriers [31, 66]. In bulk SCs, there are many energy states available that forms a nearly continuous band of states with a gap between the valence and conduction bands. If an electron in the valence band is provided with enough energy it is promoted to the conduction band and leaves a hole in the valence band. The electron-hole pair is known as an exciton, and typical separation between them is around 1–10 nm [67]. If the size of the SC crystal is of these dimensions (or less than the Bohr radius) then the exciton is confined [67]. Quantum confinement occurs when the excitons are spatially confined in one

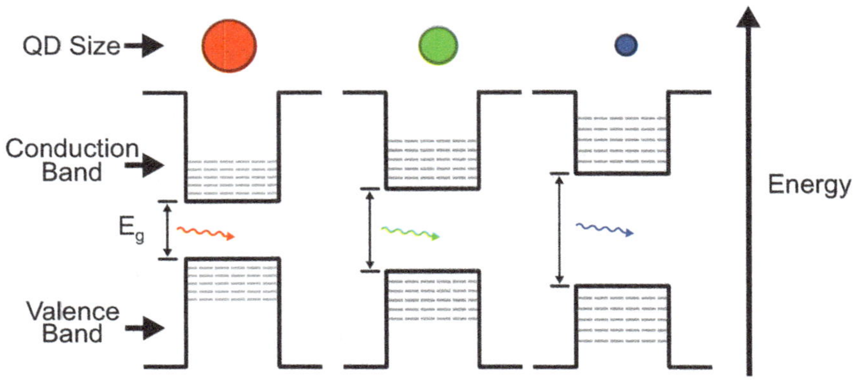

Fig. 1.3 An energy level diagram to demonstrate how energy levels become quantised for a QD (indicated by dashed lines) and how the bandgap E_g varies with size of QD. The size of QD is represented by coloured circles at the top of the diagram and QD size decreases towards the right. The coloured wavy lines represent how the wavelength and so colour of emitted photons can change with size of QD

dimension (quantum well (QW)), two dimensions (quantum wire) or 3 dimensions (QD). Quantum confinement can approximately be described by a particle in a box type description, where energy levels become quantised. The bandgap of a quantum confined system varies with size of structure as $1/d^n$ with d diameter [68–70]. The bandgap in a SC nanostructure increases as structure size decreases, and so both light absorption and emission can be tuned by control of nanostructure size [71]. A diagram that demonstrates the energy levels and bandgaps for different sizes of a QD structure is shown in Fig. 1.3. The dotted lines indicate quantised energy levels, and it is shown that for a smaller QD size (indicated by the size of the coloured circles) the bandgap E_g is larger. Bandgap and so emission wavelength can be controlled by adjusting the size of a QD.

When dealing with bandgaps in ternary SCs ($A_xB_{(1-x)}C$) an interpolation scheme is used to find the bandgap for a specific material composition. The bandgap parameter exhibits an approximately quadratic dependence on composition, and the parameter $T_{A_xB_{(1-x)}C}$ can be approximated by [72]

$$\begin{aligned} T_{A_xB_{(1-x)}C} &= xB_{AC} + (1-x)B_{BC} + C_{A-B}x(1-x) \\ &= a + bx + cx^2 \end{aligned} \tag{1.1}$$

where $a = B_{BC}$, $b = B_{AC} - B_{BC} + C_{A\text{-}B}$ and $c = -C_{A-B}$ which is called the bowing parameter. Relevant parameters for the GaAsP system are given in Table 1.1, and the bowing parameter c for calculating bandgap of GaAsP is 0.19 [73]. The bandgap E_0 the $GaAs_{(1-x)}P_x$ is then described by

$$E_0(\text{eV}) = 1.43 + 1.14x + 0.19x^2 \tag{1.2}$$

Table 1.1 Material bandgaps [72]

Material	Bandgap E_0 (eV)
GaAs	1.43
GaP	2.76

where effects of strain have not been included.

It was seen in Sect. 1.2.2 that it is possible to switch between growth of ZB and WZ structure in NWs. These different crystal phases possess different electronic properties [74]. An example of this is seen in GaP NWs, where in the WZ phase it has a direct bandgap [75] while in ZB phase it has an indirect bandgap [76]. This allows for fabrication of functional structures without changing composition, and examples of this include InP [77] and GaAs [78] single QD devices.

NWs have been shown to have some improved thermal properties [79]. Li et al. showed Si NWs had lower thermal conductivity compared to bulk [80], which is a useful property for making more efficient power generators that use thermoelectric materials [81].

With a high refractive index and a small surface roughness NWs can be ideal for use as waveguides in light-emitting devices, lasers, and sensors to name a few. A single SC NW with relatively high refractive index compared to the environment is a natural Fabry-Pérot cavity as light reflects at the end faces [52]. Waveguiding properties can be enhanced further by adding low refractive index shells to improve photon confinement and promote waveguiding inside the NW core [82]. The high refractive index and the subwavelength diameter of NWs means the electromagnetic modes of the NW tend to be leaky and interact more with the outside media [83]. This can enlarge the NW absorption cross-section to be bigger than the NW [84]. The leaky mode resonance can be tuned to a desired wavelength by changing NW diameter, and so allows for wavelength-selective absorption enhancement. The light trapping ability of NWs is superior to that of thin films [11]. Because of the antenna effect [85], NW light emitting diodes (LEDs) can have an enhanced light extraction [11, 86]. The wavelengths available in NW LEDs are able to cover the 'green-yellow' 550–590 nm range that is inaccessible using thin film technology. One way in which bandgaps can be engineered in NWs is by using strain [87–89].

The small size and high surface to volume ratio provides some useful mechanical properties, including high stiffness and high strength [10]. Surface effects play an important role in NW mechanical properties, and can either enhance or diminish them [90]. The difference in size effects on mechanical properties is mostly due to the reduction in the number of defects in its volume, with this being the reason for some NWs having ultrahigh strength. In NWs with twin boundaries, these boundaries can block motion of defects in the NW when bending and can strengthen the NW [91]. In GaAs NWs it was seen that planar defects strengthen the NW where a high density of planar defects hinders crack initiation and so strengthens the NW [92].

1.2.4 Applications

The list of potential NW applications is a long one. Many of the uses benefit from properties of NWs and many show better performance than bulk or thin film counterparts. A selection of NW applications is given here to demonstrate the potential for uses of NWs. For even more information on NW applications a range of recent reviews are available [8, 9, 11].

- NWs are a candidate for further downscaling of electronics [12, 13]. Some example uses in electronics include NW p-n diodes [93], NW field effect transistors (FETs) [94–96], logic gates [97], transistors [98], and even single electron transistors [99].
- III–V SCs can be integrated with Si. Examples of this include Si NWs with sections of optically active GaP grown axially into the NW structure [100].
- NWs can be used as a light source, with examples of LEDs [11, 36], lasers [101] and quantum cascade lasers [94, 102]. It has been shown it is possible to obtain tunable optical emission from QDs [86, 103, 104] and it is possible to produce single QD LEDs in NWs [105]. There have also been examples of integrating GaP on Si for non-linear photonics [106]. A single NW spectrometer has been fabricated [107], and NWs have been used in photodetectors [108, 109].
- Quantum devices—It has been recently shown that majorana zero-modes can exist in a NW and there is potential for future experiments that lead to topological quantum computing [110]. Examples have also been seen where superconductor/SC hybrid devices can be fabricated [111, 112]. Superconductor/SC hybrid topological networks which have potential applications as qubits for quantum computing have been recently fabricated [113].
- NW based memory [114, 115].
- Energy storage and conversion [15].
- Chemical and biological sensing [116, 117], and nano-bio interfaces [118].
- Flexible electronics—NWs offer the ability to assemble functional devices onto flexible plastic or glass substrates [119].
- Transparent electronics [120].

1.3 Defects in the III–V System

The previous sections have introduced the general topic of NWs and some properties they possess. The work presented in the results chapters deal with defects that have been observed in NWs, and so this section looks to introduce the topic of defects in NWs. Before looking specifically at defects found in NWs it is useful to first look at the types of defects that can be found in III–V systems. III–V systems are of interest because of their potential for band structure engineering in device applications through the combination of different III–V materials. The $GaAs_{1-x}P_x$ system in particular is of interest because of the range of bandgaps that can be covered by the material system by varying P content [11]. An extensive review on the use of

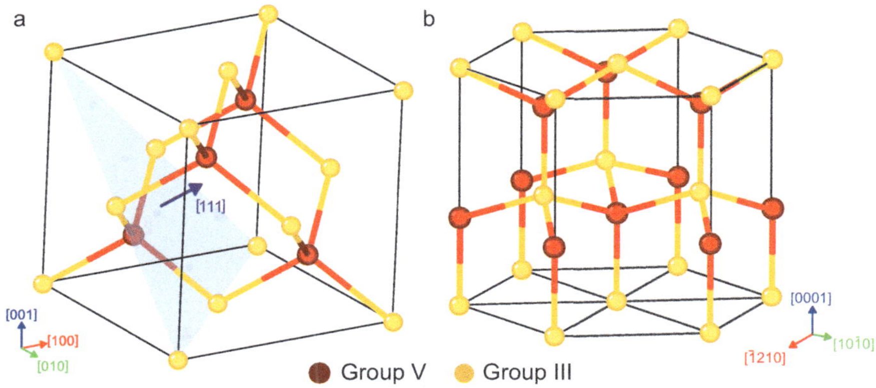

Fig. 1.4 **a** ZB crystal structure. A blue plane indicates the (111) plane with the [111] direction labelled. **b** WZ crystal structure. Yellow atoms represent the group III element and red atoms represent the group V element. Axes at the bottom indicate crystal directions for each structure

different III–V materials for NW device applications and their benefits can be found in the review by Zhang et al. [11].

Most III–V NWs can adopt the ZB or the WZ structure [11, 121, 122]. ZB is a cubic structure where two atom types form two face-centred cubic (FCC) lattices, as shown in Fig. 1.4a. The two colours represent each atom species, with yellow being group III atoms and red group V atoms. The red atoms are displaced relative to the yellow atoms by $\left(\frac{1}{4}, \frac{1}{4}, \frac{1}{4}\right)$ [123]. ZB belongs to the space group $F\bar{4}3m$ in Hermann-Mauguin notation. WZ is a hexagonal structure where each individual atom type forms a hexagonal close packed (HCP) sublattice as shown in Fig. 1.4b, and it belongs to the $P6_3mc$ space group. The ZB structure is sterically (related to spatial arrangement of atoms) more favourable, while WZ is electrostatically more favourable [124]. The ZB structure can be thought of as a polytype with a repeating $ABCABC$ stacking of layers, where the letters represent planes of atoms in the [111] direction. Similarly WZ can be considered to have a repeating $ABAB$ stacking of layers. Since in this work the ZB structure is the most common structure observed, it is useful to look at some of the crystallographic properties. When thinking about how this crystal structure can be studied, there are a few commonly used crystal directions along which the structure is viewed. The first one to consider is a ⟨111⟩ direction. The (111) plane and corresponding direction is shown by the blue plane and purple arrow in Fig. 1.4. If the crystal is viewed along this direction, it looks like the projection shown in Fig. 1.5a and the corresponding diffraction pattern is shown in Fig. 1.5b.

Another direction of interest is ⟨112⟩, which is the direction of NW vertices. In this direction the stacking sequence $ABCABC$ cannot be identified. Viewing NWs along a ⟨112⟩ direction therefore means that typical defects found in NWs do not change structure appearance [122]. An example of the structure viewed along a ⟨112⟩

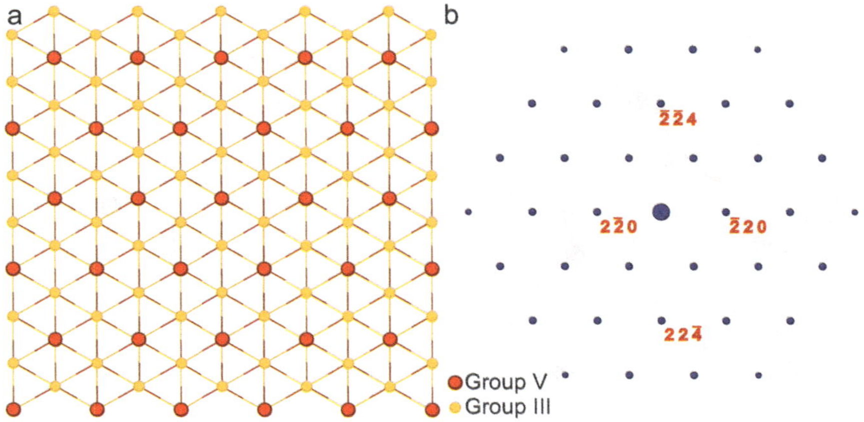

Fig. 1.5 Viewing a ZB crystal along a ⟨111⟩ direction. **a** Crystal structure **b** diffraction pattern

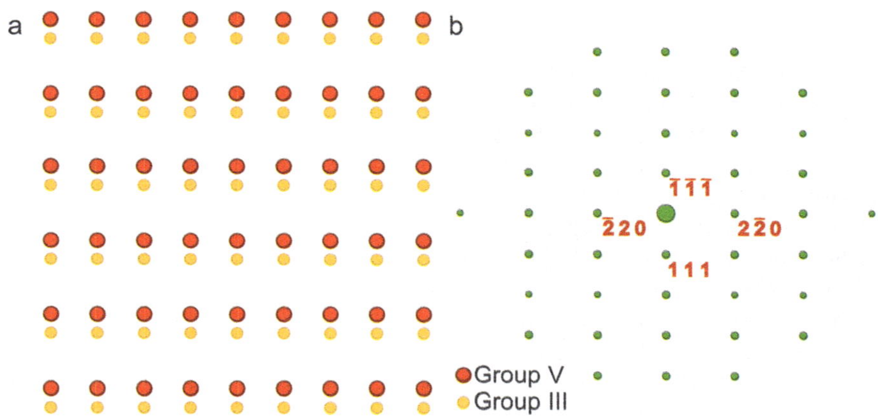

Fig. 1.6 Viewing a ZB crystal structure along a ⟨112⟩ direction. **a** Crystal structure **b** diffraction pattern

direction is shown in Fig. 1.6a and the corresponding diffraction pattern is shown in Fig. 1.6b.

The final direction to be introduced here is ⟨110⟩. This direction is important for NWs as these are normally the planes that form the vertical facets of the NW. It is also the direction in which the majority of defects in crystal structure can be observed and analysed. NWs typically have 6 {110} facets along which the NW can be viewed. The appearance of the crystal along this direction is shown at the top of Fig. 1.7a, and the corresponding diffraction pattern is shown in Fig. 1.7b.

One of the most common defects seen in III–V materials is twinning [125]. Twinning is when two crystals with the same lattice and composition meet with a sym-

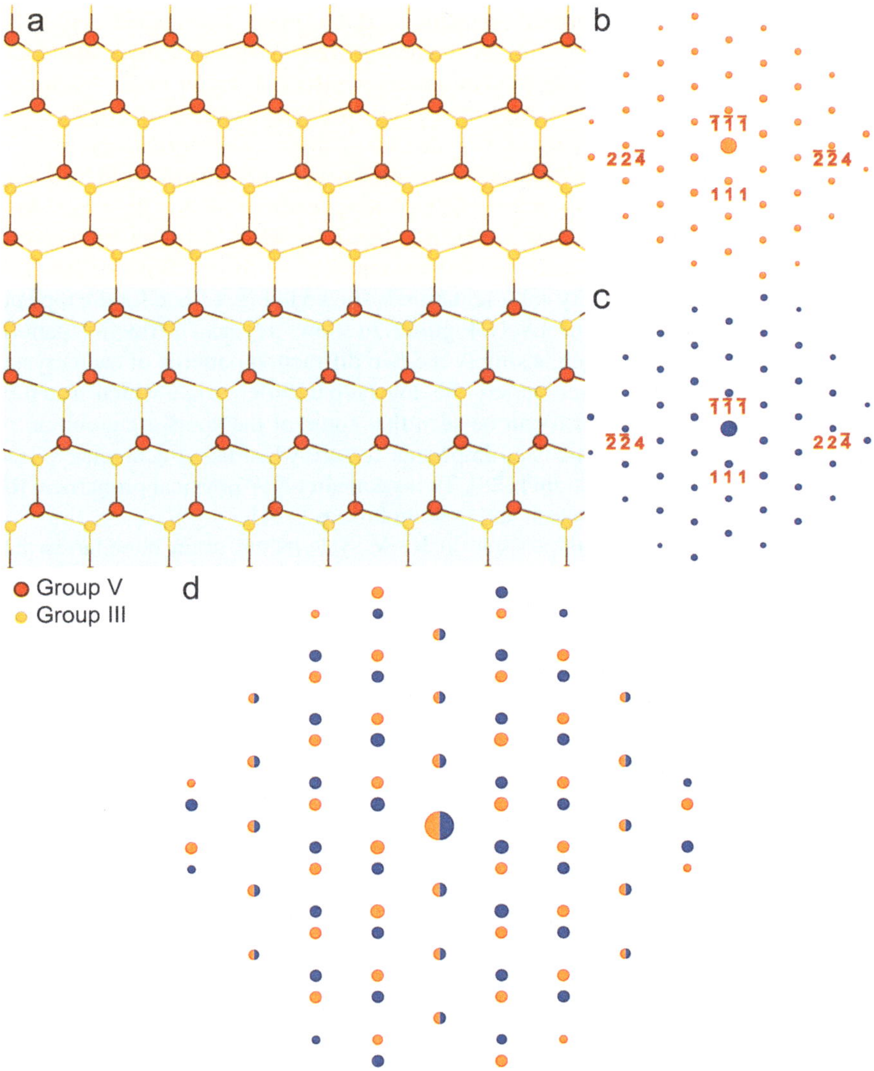

Fig. 1.7 Viewing a twinned ZB crystal structure along the $[1\bar{1}0]$ direction. **a** Crystal structure with the background coloured yellow and blue according to the crystal orientation. **b** and **c** show the corresponding diffraction patterns for the two crystal orientations in (**a**), and **d** shows the diffraction pattern of the complete twinned structure. The spots are coloured according to which crystal contributed to the spot, with some coming from both

metric relation between the crystal orientations. If the lattice planes match across the boundary, it is coherent, otherwise it is incoherent [126]. In heteroatomic cubic SCs, coherent twins can be further divided into two categories: 1 - ortho-twin (rotational) and 2 - para-twin (mirror) [121]. Across an ortho-twin boundary, polarity of the crystal structure is conserved. In ZB or WZ structures, ortho-twins correspond to 180° rotations about the [111] growth axis [127], while para-twins are produced by a mirror on the {111} plane. An example of the typical appearance of an ortho-twin when viewed along the $\left[1\bar{1}0\right]$ direction is shown in Fig. 1.7a, where a yellow/blue colour scheme in the background makes distinguishing between orientations easier. The diffraction pattern for each crystal orientation is shown in Fig. 1.7b, c for the top and bottom section of crystal respectively. Figure 1.7d shows a typical diffraction pattern for a twinned structure, which is simply the two diffraction patterns of each crystal orientation overlapped. The colours of the dots help to differentiate which structure contributes to the spots, and it can be seen that some of the spots are common to both crystal orientations. This is an important feature when using geometric phase analysis (GPA) to measure strain [128]. Twins can alter NW physical properties; for example they can reduce electron and hole mobilities [124].

A similar class of defects possible in III–V systems are grain boundaries that separate two crystals with different orientations [129, 130]. They can be further divided into small and large-angle grain boundaries depending on the angle between the two crystal planes that meet to create the boundary. One way in which a grain boundary can be described is using what is known as a coincidence site lattice (CSL). If the two misoriented crystal lattices are superimposed, a number of lattice sites are coincident (one in Σ) and form their own 3D super-lattice, known as a CSL. The ratio of CSL sites to lattice sites of one of the misoriented crystals is described by $\frac{1}{\Sigma}$ [131]. The CSL does not itself define an interface, though an interface is likely to have a low energy if Σ is small.

Another type of defects that can be common in III–V materials are dislocations. Dislocations are lines of structural discontinuity, and allow material to deform without destroying the basic crystal structure. Dislocations either form a closed loop inside the crystal or they extend to the crystal surface [131]. A dislocation can be characterised by the Burgers vector b and describes the magnitude and direction of lattice distortion. The Burgers vector can be found by drawing a Burgers circuit around a dislocation. To do this, a series of lattice vectors is taken clockwise around the dislocation to form a loop. This path is then drawn in the perfect lattice, where it does not form a closed loop. The vector required to close the loop defines b for the dislocation. This is illustrated using a simple example in Fig. 1.8. The circuit in the perfect lattice starts at S and finishes at F, and the yellow vector b is required to close the loop.

The value of b is determined by the crystal structure, and is usually equal to the shortest lattice vector available [131]. In FCC structures, the shortest lattice vector has components $a/2\ [110]$ where a is the lattice parameter, and the vector magnitude is $a/\sqrt{2}$.

In a typical crystal, the core radius of a dislocation is $\approx$0.25 nm and a typical outer radius of the associated strain is $\approx$2.5 µm [130]. Energy of an edge dislocation

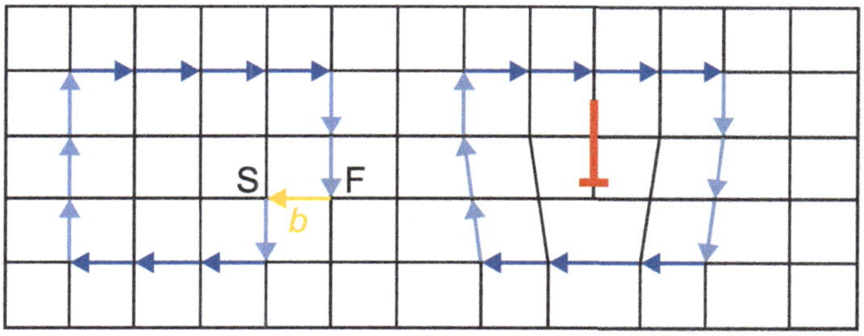

Fig. 1.8 Using a Burgers circuit to find the Burgers vector b (indicated by a yellow arrow) of a dislocation. The circuit in the perfect lattice starts at S and finishes at F

is approximately Gb^2 per unit length of dislocation, where G is shear modulus. The strain field of a dislocation having a long-range character means that dislocations influence the behaviour of other dislocations. Dislocations of opposite sign will attract each other while same sign dislocations will repel each other. For two edge dislocations on the same slip plane the force between them is of the form [130]

$$F = \frac{Gb^2}{(1-\nu)2\pi r} \tag{1.3}$$

where ν is Poisson's ratio, and r is distance between the dislocations. A dislocation may also experience an image force if it is sufficiently close to a surface. The image force on an edge dislocation is described by [132]

$$F = -\frac{Gb^2}{4\pi\,(1-\nu)\,d} \tag{1.4}$$

where d is the distance from the surface. This becomes more complicated when applied to the NW system, as shown by Ye et al., once the full 3D shape of the NW and surface stress effects are considered [133].

As was seen earlier, the FCC structure can be described by the stacking of $\{111\}$ planes in the sequence $ABCABC$, and a top down view of an A plane is shown in Fig. 1.9a.

The yellow circles represent the A layer, and positions are labelled for layers B (red) and C (blue). The $\{111\}$ planes in the FCC system can slip across each other. If atoms of the B plane, labelled in their starting position $B1$ in Fig. 1.9a, moves to the next closest B position, $B2$ via the vector b_1, then this leaves a perfect crystal and this describes a perfect dislocation with a Burgers vector the same as a lattice vector

$$\mathbf{b} = \frac{a}{2}\langle 110\rangle \tag{1.5}$$

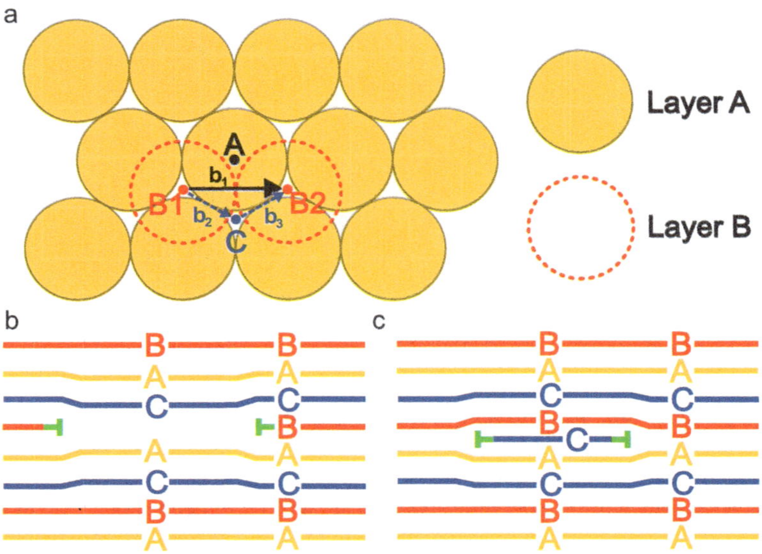

Fig. 1.9 **a** Stacking of {111} planes in FCC structures and slip of {111} planes. **b** An intrinsic stacking fault. **c** An extrinsic stacking fault

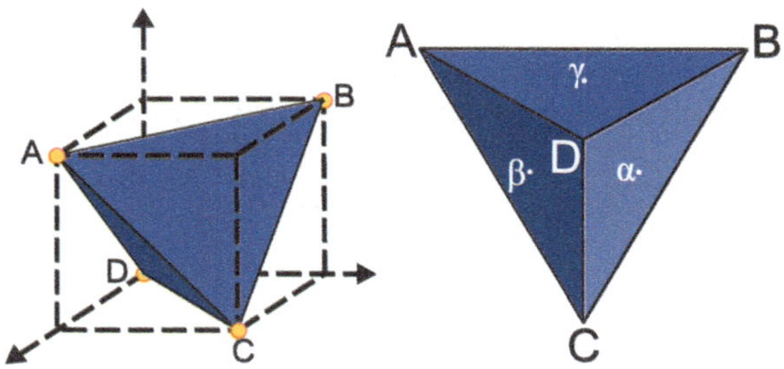

Fig. 1.10 Thompson's tetrahedron used to describe defects in FCC structures. The letters A, B, C and D correspond to corners of the tetrahedron and the midpoints of the opposite faces are denoted by α, β, γ and δ. The δ face is not visible on the tetrahedron shown on the right

So far the defects considered have lattice Burgers vectors. Another class of defects, known as partial dislocations, have a Burgers vector smaller than a lattice vector. It can be helpful to use what is known as the Thompson tetrahedron to describe these defects [131]. The tetrahedron is formed using the FCC structure, with the vertices ABCD as shown in Fig. 1.10. The planes correspond to {111} planes and the edges to $\langle 110 \rangle$ directions. The midpoints of the faces are labelled $\alpha, \beta, \gamma, \delta$. A perfect dislocation can be described using this system by, for example, AB. It is more energetically favourable for this transition to occur in two steps by taking the path $\mathbf{b}_1 = \mathbf{b}_2 + \mathbf{b}_3$ in Fig. 1.9a, or using the Thompson tetrahedron, $AB = A\gamma + \gamma B$. The perfect dislocation is split or "dissociated" into two partial dislocations with Burgers vectors $\mathbf{b}_2$ and $\mathbf{b}_3$

$$\frac{a}{2}[110] \rightarrow \frac{a}{6}[211] + \frac{a}{6}\left[12\bar{1}\right] \tag{1.6}$$

These partial dislocations are called Shockley partial dislocations, and their Burgers vector lies in the plane of the fault [130]. Generally the pair of partials that form repel each other and there is usually a stacking fault between them, with the stacking changing from $ABCABC$ to $ABCACABC$. Shockley partials are glissile, meaning they can move by conservative glide motion [126].

Two perfect dislocations can meet along a slip plane, and each perfect dislocation can split into two Shockley partials; a leading and a trailing. When two leading partials combine, they form a different dislocation with a Burgers vector not in the plane of the fault, and is known as the Lomer-Cottrell lock (LCL). These have a Burgers vector of the form [131]

$$\mathbf{b} = \frac{a}{6}\langle 110\rangle \tag{1.7}$$

In Thompson tetrahedron notation, this can be represented by $DA \rightarrow D\beta + \beta A$, $BD \rightarrow B\alpha + \alpha D$ and when the two Shockley partials αD and $D\beta$ interact, then a lock $\alpha\beta$ is formed [131]. Another type of lock can form when two perfect dislocations with perpendicular Burgers vectors glide on intersecting planes. The reaction results in a sessile Hirth lock of the form [131, 134]

$$\mathbf{b} = \frac{a}{3}\langle 001\rangle \tag{1.8}$$

Another type of partial dislocation is the Frank partial, and these outline stacking faults that are formed by inserting or removing a region of a $\{111\}$ plane. Figure 1.9b, c show an intrinsic stacking fault and an extrinsic stacking fault from a loop of Frank partials respectively. Frank partials have a Burgers vector of the form [131]

$$\mathbf{b} = \frac{a}{3}\langle 111\rangle \tag{1.9}$$

Frank partials are represented by $A\alpha$, $B\beta$, $C\gamma$ etc. using the Thompson tetrahedron notation. Frank partials are sessile [126].

Whilst examples of the defects introduced in this section found in NWs are examined in Chaps. 3 and 4 using atomic resolution electron microscopy, more general overviews on the use of electron microscopy to study defect structures can be found in some recent reviews [126, 135, 136].

1.3.1 Defects in Nanowires

This section looks more closely at imperfections that occur specifically in NWs. First, to better understand what parameters are important in defect formation in NWs, the critical radius of dislocation formation is considered in more detail.

Glas [137] showed that there is a radius-dependent critical thickness below which no dislocations should form in a NW. This means that for each material system, there exists a critical radius below which NWs can grow free of dislocations. One reason for this behaviour is NWs have free surfaces at their sidewalls, not just at the tip, leading to more efficient strain relaxation. Another reason for this behaviour is because of surface image forces. In order to find a value for the critical radius, the energy of the NW system with and without dislocations can be found using an approach similar to that used in calculations of 2D layers.

The formation of dislocations reduces the misfit between substrate and layer and reduces elastic energy of the system, however dislocations have their own strain field and core energy which increases the total energy of the system. The elastic energy of a dislocation parallel to a free surface can be calculated as a function of distance from the surface. The energy per unit length of a dislocation (w_d) is given by [137]

$$w_d = \frac{E(1 - \nu \cos^2 \theta)b^2}{8\pi(1 - \nu^2)} \left(1 + \ln \frac{\bar{h}}{b}\right) \tag{1.10}$$

where E is the Young's modulus, ν is Poisson's ratio, θ is the angle between the dislocation line and its burger vector, b is the core cutoff radius for the calculation of the elastic energy and $\bar{h}$ is effective distance to the surface. Glas makes the assumption that $\nu = 1/3$, and the same assumption is applied in this chapter. In order to get an expression for critical radius, the difference in energy ΔW between a NW interface with a dislocation pair and a fully coherent interface can be used. If $\Delta W > 0$ the interface is stable, if $\Delta W < 0$ it is favourable to introduce dislocations. $\Delta W(r_0, h_c) = 0$ then defines the critical thickness h_c as a function of radius r_0. Critical radius can be found from the solution of [137]

$$\frac{2\pi}{A_v} \left(\frac{\alpha^2 b_{\text{eff}}^2}{4} - \alpha b_{\text{eff}} \epsilon_0 r_0^c\right) + C \left(1 + \ln \frac{\beta r_0^c}{b}\right) = 0 \tag{1.11}$$

$$C = \frac{(1 - v \cos^2 \theta)b^2}{2\pi(1 + v)} \tag{1.12}$$

where $A_\nu = 27.3 \pm 0.55$ for $\nu = \frac{1}{3}$, $\alpha = \frac{4}{\pi}$, ϵ_0 is the relative difference of lattice parameter between two materials, and $\beta = \frac{2}{\pi}$. Through different θ values, the critical thickness can be calculated for different types of dislocations. For pure edge dislocation ($\theta = \frac{\pi}{2}$, $b_{\text{eff}} = b$) and for 30° partial dislocations ($\theta = \frac{\pi}{6}$, $b_{\text{eff}} = \frac{\sqrt{3}}{2}b$).

As a reference for the system from which the results are obtained in the results chapters, a calculation is made here to find the effective lattice constant a, the strain ϵ and finally critical radius r_c for $GaAs_{(1-x)}P_x$ for different P compositions x. The calculation is performed using Si as the substrate, and a linear interpolation is used to get values of the lattice constant for different P content, as given by Vegard's law [139]

$$a_{\text{GaAs}_{(1-x)}\text{P}_x} = (1 - x)\, a_{\text{GaAs}} + x a_{\text{GaP}} \tag{1.13}$$

Table 1.2 Material lattice constants [138]

Material	a(Å)
Si	5.4310
GaAs	5.6536
GaP	5.4506

Plots of effective lattice constant and strain using the lattice constants given in Table 1.2 are shown in Fig. 1.11a. These are used to solve Eq. 1.11 to find critical radius r_c for 30° partials and edge dislocations, and are shown in Fig. 1.11b.

One of the most common imperfections found in NWs is twinning. It was seen relatively early in the field that twins can be very common in $\langle 111 \rangle B$ III–V NWs [13]. Twins can negatively affect optoelectronic performance and enhance charge recombination [13, 140], and so understanding how they form in NWs is useful to know. By using Glas' nucleation model [59] and considering the change in Gibbs energy when a nucleation of twinned material occurs, Yuan et al. showed that the probability of forming twin-free (p_{TF}) and twin nuclei (p_{Twin}) is an exponential function of the nucleation energy barrier [55]

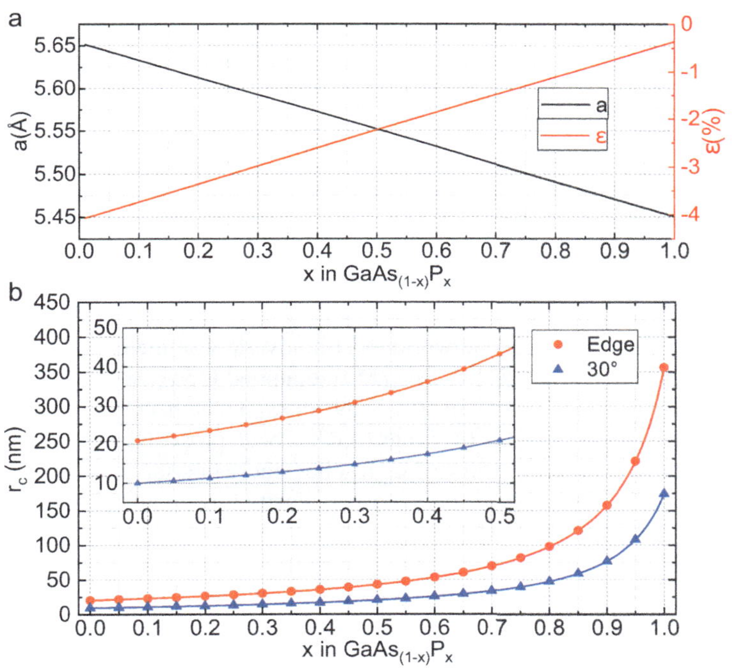

Fig. 1.11 **a** Lattice constant (black) and strain (red) as a function of P composition in GaAsP. **b** Critical radius of NW for different types of dislocations as a function of P composition in GaAsP. The inset shows a closer view of the first half of the plots

$$\frac{p_{\mathrm{TF}}}{p_{\mathrm{Twin}}} = \exp\left(\pi h^2\left(\frac{\Gamma_{\mathrm{WZ}}^2}{\frac{h\Delta\mu}{\Omega} - \phi} - \frac{\Gamma_{\mathrm{ZB}}^2}{\frac{h\Delta\mu}{\Omega}}\right)\right) \tag{1.14}$$

where h is the height of a monolayer (ML) in the $\langle 111\rangle$ direction, Γ is lateral surface energy, $\Delta\mu$ is the chemical potential in the liquid, Ω is the elementary volume of GaAs, ϕ is 0 for a ZB nucleus and $0.023\,\mathrm{J\,m^{-2}}$ for a WZ nucleus which is half the stacking fault energy in GaAs [141]. Twin formation probability falls with increased wetting angle β, and also depends on growth temperature and As concentration [142].

Twin boundaries can form in any $\langle 111\rangle$ direction other than growth direction. Inclined twins often lead to kinks in NWs [136]. When imaging twinned materials, inclined twins present that are not parallel to beam direction can cause additional peaks of intensity in between dumbbells [136, 143]. In NW structures where there are sections of polytype/twinned material that overlap in the direction of the electron beam, the overlap can be identified in diffraction patterns (DPs), with double diffraction from twinned crystals and forbidden reflections occurring in some cases [144, 145]. The superposition of two crystal phases has been reported to produce extra $\frac{1}{3}$ (111) spots which results from diffraction from the first crystal phase and subsequent diffraction again from the second crystal phase [146].

While twins in NWs are most commonly across the growth direction $\left[\bar{1}\bar{1}\bar{1}\right]$, other twins have also been observed in NWs. It was shown by Sanchez et al. that $\Sigma3\{112\}$ twins can be found in self-catalysed GaAsP NWs grown using solid source MBE [1], and they have also been reported in GaAs NWs [147]. It was shown that these {112} twins can act as steps in {111} twins, and since polarity was conserved across the {111} twin, this means the {112} twin is an uncommonly seen para-twin. These steps appear to obey a multiple of 3 rule, with the height of the step always existing in multiples of 3, and examples with heights of 3 and 6 were observed. An example of this type of twin interface was also seen where there was a section of twinned material, with two {111} twins and a {112} twin. {112} twins have topological properties similar to those of other line defects, and they must either form closed loops or continue to the surface [1, 129].

This $\Sigma3\{112\}$ interface is known to exist in a variety of materials including Cu [148, 149], Au [150], Al [151], β-SiC [152], diamond [153], Si [154–157], III–Vs [158], and II–VIs [159, 160]. Feng et al. [159] showed that this type of boundary can create deep gap states, and this twin interface has been proposed to be a preferential site for impurity atoms [157, 159–161]. The effect these interfaces have on electronic properties in GaAs NWs was examined using density functional theory (DFT) [162, 163]. It was shown the band gap was effectively closed by midgap states [1], and so suggests the steps act as charge recombination sites which would influence opto-electronic properties. States appearing in the bandgap as a result of {112} boundaries have been seen before in other material systems, e.g. in diamond [153] and Ge [164], meaning this property is not unique to the GaAs system.

Grain boundaries are briefly mentioned here as these have been seen in some NWs, with examples including gold NWs [165] and copper [166]. It was shown that grain boundaries can act as preferential sites for incorporation of catalyst material

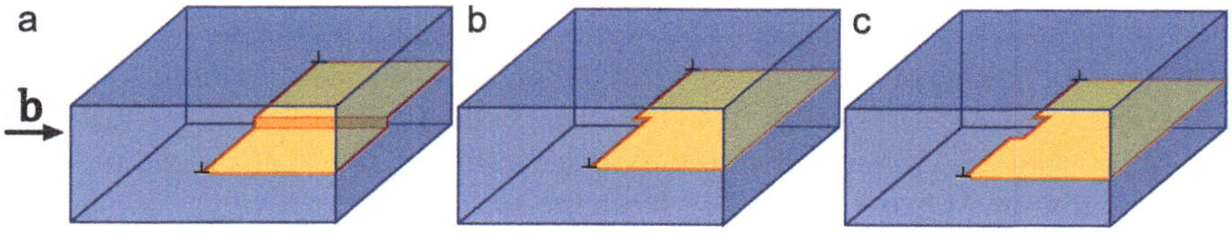

Fig. 1.12 Mechanisms of dislocation motion. **a** A dislocation moving to another slip plane by climb forms a jog. Dislocation glide through movement of kinks along dislocation core via **b** single kink or **c** double kink nucleation

in Au catalysed NW growth of Si NWs [167, 168]. Incorporation of Au in this way was shown to have detrimental effects on NW properties [169].

Another type of defect that have been reported in NWs are point defects [136]. Some examples of these include gold impurities in Si NWs [170], and it was shown using atomic scale scanning tunneling microscopy (STM) that point defects can be found in the surface of InAs and GaAs NWs [171].

1.3.2 Mechanisms of Defect Motion

Dislocation motion in SCs can be a thermally activated process [172]. There are two main mechanisms through which a dislocation can move and they fall under the categories of conservative and non-conservative motion.

Non-conservative motion, or climb, occurs when motion is normal to the Burgers vector [131]. Climb requires thermal activation, and the most common climb process involves diffusion of vacancies towards or away from the dislocation. Climb of a short section of a dislocation line results in the formation of jogs. Jogs are steps on the dislocation which move it from one atomic slip plane to another (Fig. 1.12a). Steps which displace a dislocation on the same slip plane are kinks (Fig. 1.12b, c). Jogs and kinks are short elements of a dislocation with the same Burgers vector as the dislocation line they displace.

Conservative motion, or glide, occurs when motion is in the plane that contains both the dislocation line and Burgers vector, and dislocations that move in this way are glissile. Dislocations that can not move in this way are sessile [131]. Glide of dislocations occurs such that the distortion associated with it is minimal [173], and is normally over the most densely packed planes of a lattice. During glide, atoms near the dislocation shift and effectively move the extra half-plane of atoms. Dislocation glide in SCs can be limited by the Peierls mechanism [174], and takes place by movement of thermally generated kinks along the core. For motion to occur, the dislocation must move from one Peierls potential valley to the next. If a section of a dislocation moves to an adjacent potential valley, the part that connects the moved section of the dislocation to the original dislocation valley is a kink. Each kink displaces the defect by one lattice translation vector on its glide plane [175].

A single kink is shown in Fig. 1.12b, which can nucleate at the crystal surface and subsequently propagate along the dislocation line. If dislocation motion is initiated along the dislocation line away from the surface, then a double kink nucleates, as shown in Fig. 1.12c. After a double kink nucleation, the two kinks move in opposite directions and separate until they reach the surface. Alternatively, if kinks have nucleated elsewhere along the dislocation line, then kinks of opposite direction can meet and annihilate. For long dislocations, the dislocation velocity is controlled by the kink-pair nucleation rate per unit length of dislocation, and kinks travel along the dislocation until they collide with a kink of opposite sign and annihilate [123]. For short dislocations, kinks are more likely to reach the surface before meeting a kink of opposite sign [123]. In this case the probability of kink-pair nucleation and hence velocity is proportional to the length of the defect. Dislocation glide in strained layers was shown to have a linear dependence of velocity on crystal thickness [174]. Generally speaking, the velocity of a dislocation is described by

$$v = JLd \tag{1.15}$$

where v is velocity, J is double-kink formation rate per unit length of dislocation, L length of dislocation and d distance between adjacent Peierls valleys. In the case of no-collisions of kinks, the velocity can be described by an equaiton of the form [174]

$$v = cLF \exp\left(-\frac{Q}{k_B T}\right) \tag{1.16}$$

where c is a constant, F is force, Q is activation energy, k_B is Boltzmann constant and T is temperature.

Edge dislocations move by glide, limited to a specific plane, and at higher temperatures they can move by climb. Partial dislocation motion is dictated by what plane the Burgers vector and dislocation line are in. In the FCC system, Shockley partials are glissile while Frank partials are sessile, meaning Frank partials do not move by conservative motion [130]. For a kink to move one lattice translation along a partial dislocation in the ZB structure, a single atomic bond must be broken and reformed, and the energy barrier for this to take place determines the activation energy Q for dislocation movement.

The motion of dislocations can be influenced by a variety of factors. Impurities in a crystal can react with a dislocation and in general impurity atoms segregate to dislocation cores and this results in the pinning of the dislocation which becomes immobilised [172, 176, 177]. Any changes that occur to the crystal surface can alter dislocation behaviour [178]. When dealing with III–V–V materials the exact composition may influence dynamics, and to illustrate how much this can affect dynamics it was seen that defect velocity in GaP was 100 times that in GaAs [179]. Interaction of an electron beam is also known to influence defect dynamics [180–182].

With the $\Sigma3\{112\}$ twins being the most commonly seen in the NWs investigated in this work, a note here is made about how these type of defects are expected to move. The core structure of these {112} twins contains the structural motifs of $\frac{1}{6}\langle 112\rangle$ intrinsic and extrinsic partial dislocations, and can be considered to be a dislocation dipole made up from a pair of these. Similar to a dissociated dislocation, an attractive force between the two partials prevents independent movement, and both partials move with conservative glide motion. Motion of these {112} twin facets has been previously seen and reported by Xu et al. [148]. The study of how these defects move can help to provide a better understanding of how these defects can be removed from NW systems. These types of defects are detrimental to NW performance and so their removal from NWs is desirable.

1.4 Nanowire Heterostructures

A NW heterostructure is a combination of two or more different materials within the same NW structure [12]. The most common varieties of heterostructures in NWs are axial and radial, with heterostructures created when the composition of the NW is modulated along the axis or radius of the NW respectively [183]. With high quality growth, interactions of low-dimensional components and their interfaces can result in electronic, photonic, magnetic, and thermal characteristics that are superior to those of planar geometries [183]. Typical axial structures include QDs and typical radial structures include QWs, and core/shell structures. While this section mostly looks at axial heterostructures, some radial heterostructures are worth mentioning. The shell in core/shell structures can act as a passivation layer that reduces the impact of undesirable effects of surface states and improve device performance [9, 94]. Defects that are normally found at surfaces reduce the lifetime of minority carriers [184, 185], and passivation shells prevent these defects from being active. Radial growth can also be used to produce QWs and superlattices [186].

For heterostructures in NWs, the interface is important and can play an important part in device function [183, 187]. The sudden termination of a material at an interface has a few physical consequences. The interruption of translational symmetry of the lattice leads to new states that are not part of the bulk band structure and can alter electrostatic potential [8]. Such a change in electrostatic potential can act as a barrier to charge transport. Interfaces can lead to the scattering of electrons, photons, phonons and can be used to allow or prevent the transport of electrons and photons in an energy and spin-selective manner [183]. Different types of interfaces are possible in NWs, and are used for example in *p-n* junctions [93].

Growth of heterostructures is possible via a few techniques, with the main method being via material modulation during bottom-up growth using techniques such as chemical vapour deposition (CVD) and MBE. A variety of heterostructures are possible in NWs that would not be accessible in bulk or thin film materials because of the ability of NWs to more efficiently relieve strain/stress. An early example of the successful growth of defect-free axial heterostructures was seen in the GaAs/GaP system [186] with lattice mismatch 3%. The idea of producing a heterostructure is simple since it requires changing the source material fluxes, but in practice a few issues can become apparent. One issue is changing source fluxes can cause instabilities in the catalyst droplet and can lead to things like NWs kinking when growth conditions are not optimised [188]. Another issue can be in achieving sharp interfaces, with a common issue in VLS growth being the reservoir effect which causes a graded interface from residual atoms of a species left over in the catalyst droplet once the source for that species has been turned off. It has been shown that for III–V structures, the switching of the group V element results in sharper heterostructures compared to those formed from switching group III composition [5, 189]. Sharp interfaces are possible using MBE growth because of good control of material fluxes and growth can be aided by the monitoring of the completion of each ML by using reflection high-energy electron diffraction (RHEED) [5]. Early examples of successful growth of sharp interfaces include InAs/InP structures grown by Björk et al. [190].

In many applications in electronic devices, the interface sharpness is an important parameter [8, 99, 191]. The interface sharpness of a heterostructure depends on the growth process and can range from atomically sharp to graded over several times the NW diameter [8, 192]. An example use of atomically sharp interfaces is in QWs and quantum-cascade lasers [5]. To illustrate how important sharp interfaces can be, an example is a resonant tunnel diode (RTD) constructed with an InAs QD between two InP barriers, embedded in an InAs NW. Sharp spectral features were observed in tunnelling spectra, which correspond to well defined quantum levels which are enabled by very sharp interfaces [191]. Sharp interfaces also enable the possibility of producing single photon sources [94].

In Chap. 5, a selection of models are fit to some experimental data from GaAs QD structures grown axially into GaAsP NWs, where interest lies in looking for trends in sharpness of QD interfaces for different radii of NW. A part of this analysis looks to compare these models and to see which models fit the experimental data most closely. The following sections introduce the different models considered, with two of the models specifically developed for the NW system and indicates which parameters influence interface sharpness.

1.4.1 Dubrovskii Kinetic Growth Model

A model developed by Dubrovskii [24] looks at describing growth of axial heterostructures in NWs via VLS. This model captures the dependence of heterostructure interface sharpness on growth parameters and NW radius using a kinetic approach. The model considers the vapour fluxes and how switching between them changes atomic species present in the catalyst droplet and how this then changes the solid composition as growth proceeds. The complete model involves a large number of parameters which are listed in Table 1.3.

Growth is dictated by change in concentration and the rate at which different pairs of atomic species bond and create new solid material at the NW/catalyst solid/liquid interface. A few different components contribute to change in species concentration, including new material entering the droplet from the source flux, from surface dif-

Table 1.3 List of parameters used in Dubrovskii's model

Parameter	Definition
a_i	Effective influx
$a_i^{(s)}$	Stationary effective influx
α	Geometrical factor
β	Contact angle
c_k	Relative atomic concentrations of species k
G	Axial growth rate
K_k	Bonding/incorporation rate
N	Total number of atoms
N_i	Number of atoms of species i
R	Nanowire radius
τ_i	Relaxation time to stationary values
t	Time
U_k	Material outgoing flux kinetic coefficient
V_k	Material influx kinetic coefficient
x	Nanowire composition
z	Nanowire height

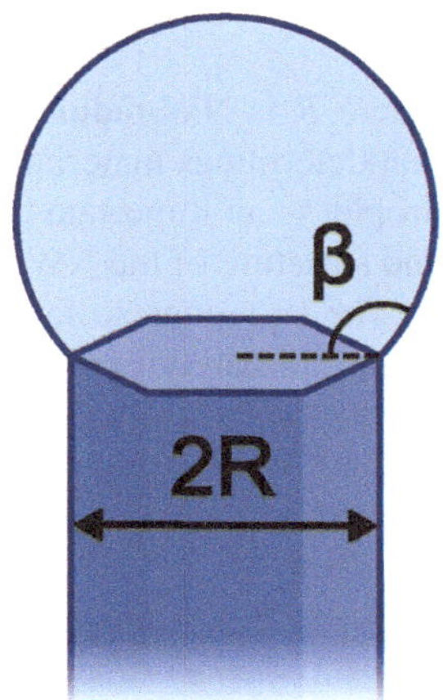

Fig. 1.13 NW geometrical parameters wetting angle β and NW radius R

fusion to and away form the droplet, and desorption of material. The geometry and size of the droplet can influence growth, with both radius R and contact angle β as labelled in Fig. 1.13 important parameters. Growth rate is then determined by the rate at which new bonds form. When there is a mix of more than two different species, the composition of new material at a given time will depend on the concentration of all relevant species in the droplet at that time, with each species having a different incorporation rate. A key parameter used in this model summarises this type of behaviour, and is the effective influx of species i, a_i

$$a_i = \frac{K_i c_i}{U_0} \tag{1.17}$$

where K_i is the bonding rate between the group III (0) species and group V (i) species, $c_i = N_i/N$ is the relative atomic concentration of species i with N_i the number of atoms in the droplet of a total of N atoms in the droplet. U_0 is the kinetic coefficient for material outgoing flux of species (0). In Ga catalysed III–V VLS growth, axial growth rate is limited by the group V fluxes [44, 193, 194].

For the case of a group V based heterostructure grown in a Ga catalysed III–V structure, than the dynamics can be described using the equations

$$\frac{\mathrm{d}a_i}{\mathrm{d}t} = \frac{1}{\tau_i}\left[a_i - a_i^{(s)}\right] \tag{1.18}$$

Which is the rate of change of a_i, where $a_i^{(s)}$ is the stationary value of a_i the system adjusts to once fluxes have changed and an equilibrium state is reached. τ_i is the relaxation time to the stationary $a_i^{(s)}$ values and is given by

$$\tau_i = \frac{R}{\alpha}\frac{1}{(K_i + U_i)} \tag{1.19}$$

where R is NW radius, α is a geometrical factor and U_i is the kinetic coefficient that determines material outgoing flux. It can be seen here that the geometry of the droplet is an important aspect of growth dynamics and ultimately the composition and structure of the NW.

The equations so far presented all describe aspects of the dynamics of growth, while ultimately interest is in composition of the final solid structure as a function of coordinate z along the NW axis. The next quantity of interest is how the height z of the NW changes with time t as it grows. This can be obtained by integrating the NW growth rate G

$$z = \int_0^t dt' G(t') \cong V_0 \int_0^t dt' \left[a_1(t') + a_2(t')\right] \tag{1.20}$$

where V_0 is a kinetic coefficient for material influx of specimen (0). This now gives the ability to describe a_i and z as functions of time. The final quantity required for a profile to be built is the composition x, which is given by the expression

$$x = \frac{a_1}{a_1 + a_2} \tag{1.21}$$

In order to build a profile, values for a_i and z must be found for a series of t values, and these can then be combined using Eq. 1.21. Expressions for $a_i(t)$ can be obtained by integrating Eq. 1.18 which gives

$$a_i(t) = a_i^{(s)} + C_i \exp\left(\frac{t}{\tau_i}\right) \quad (1.22)$$

where C_i is an integration constant. Similarly, an expression for $z(t)$ can be found

$$z(t) = V_0\left(\left(a_1^s + a_2^s\right)t + C_1\tau_1\left(\exp\left(\frac{t}{\tau_1}\right) - 1\right) + C_2\tau_2\left(\exp\left(\frac{t}{\tau_2}\right) - 1\right)\right) \quad (1.23)$$

To bring this together, the a_1 and a_2 in the expression for $z(t)$ can be replaced by values given by Eq. 1.22, and finally these values can be brought together using Eq. 1.21 to get a profile of $x(z)$. In this model τ shows that heterostructure interfaces may not be atomically sharp, and the dependence of τ on R here is where sensitivity of interface sharpness on NW radius is described by this model.

To illustrate what is happening during the growth of a heterostructure, an example hypothetical structure is considered here. Since experimentally a GaAsP-GaAs-GaAsP system is studied, this system will be used to build an example theoretical structure. For this, a choice needs to be made in how the different source material fluxes are changed for different sections of growth time. The initial section (1) will be a GaAsP section with both group V fluxes active, there will then be a period of time where the P flux is stopped and the As flux increased to compensate and maintain group V flux (2), and then there will be a final return to both group V fluxes active (3). In building the profile, these changes are reflected by changing $a_i^{(s)}$ in Eq. 1.22. For an example profile $a_{P,(1)}^{(s)} = 0.2$, $a_{As,(1)}^{(s)} = 0.8$, $a_{P,(2)}^{(s)} = 0$, $a_{As,(2)}^{(s)} = 1.0$ and $a_{P,(3)}^{(s)} = 0.2$, $a_{As,(3)}^{(s)} = 0.8$ Other parameters used are $\tau_P = -2$ $\tau_{As} = -1$, $V_0 = 1$ and $C_{1,2} = 1$. The different growth sections are given time periods $t_{(1)} = 10$ s, $t_{(2)} = 20$ s and $t_{(3)} = 20$ s. A profile is built using a time step of 0.1 s. Parameters $a_P(t)$, $a_{As}(t)$, $z(t)$ and $x(z)$ are tracked and shown in Fig. 1.14, with the data in each plot split into each time section by different colours.

1.4.2 Priante/Glas Model

This model was initially developed to describe an AlGaAs/GaAs system and is based on kinetic equations for Al entering and leaving the liquid catalyst droplet coupled with other equilibrium or kinetic considerations on the relationship between Al content in the liquid (y) and solid (x) [23, 195, 196]. The model uses thermodynamic functions for the system of interest with parameters taken from known bulk values. NW growth is represented by the addition of individual MLs each with their own composition x [195]. A key aspect of the model being the idea of decoupling between group III and V species, i.e. that the NW growth rate is set by the group V concentration in the liquid catalyst droplet. A key equation in the model is one which describes

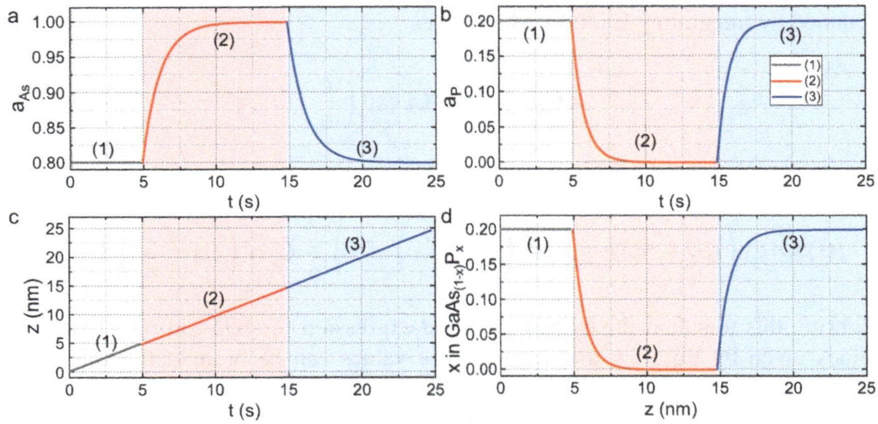

Fig. 1.14 Example simulated composition profile using Dubrovskii's growth model. Growth parameters used are given in the main body of text. The different line colours correspond to three different growth stages

the change in the amount of an atomic species in the liquid droplet [195]

$$\frac{dy_{\mathrm{Al}}}{dt} = -grx\,(y_{\mathrm{Al}}) + \frac{\phi}{N^{\mathrm{l}}} \tag{1.24}$$

where g is a NW-specific geometric factor that depends only on NW radius and droplet contact angle, r is instantaneous rate of formation of MLs, ϕ is Al atoms per unit time entering the droplet, and N^l is the total number of atoms in the liquid droplet. The incoming Al current can be obtained from experimental data, based on sample geometry and growth time. Each new ML adopts a composition dictated by the concentration in the liquid droplet before the ML was formed [195]. This depletes the liquid of the corresponding number of Al atoms (fixed by R) which then sets the new y_{Al} for the next ML. If the growth rate is assumed to be constant and using axial coordinate ζ and $d\zeta = \rho(t)hdt$ with h the height of a ML ($h = \frac{a}{\sqrt{3}}$ [197]), ρ the instantaneous NW growth rate, and i_{Al} the normalised rate of Al input into the droplet then

$$\frac{dy_{\mathrm{Al}}}{d\zeta} = -\frac{g}{h}x + \frac{i_{\mathrm{Al}}}{\rho h} \tag{1.25}$$

To obtain a compositional profile of $x(\zeta)$ a relation between composition of the liquid catalyst and the solid, $y(x)$ is needed. Two models have been developed by Glas et al. [23] to address this.

In the first model, the decoupling between III/V is total, and composition is taken as the equilibrium composition corresponding to the concentration of the minority group III in the liquid. This can be found from conditions on differences in chemical potentials $\Delta\mu_{\text{Al,As}} = 0$, $\Delta\mu_{\text{Ga,As}} = 0$, using the relevant thermodynamic equations for the chemical potentials [23].

In a separate piece of work by Dubrovskii et al. this approach was generalised to work for most material systems [196]. The generalised relation between liquid and solid concentrations is given by

$$y = \frac{x}{x + (1 - x)e^{2\omega_s\left(x - \frac{1}{2}\right) + b}} \tag{1.26}$$

where b is affinity and ω_s is interactions in solid. This reduces to the Langmuir McLean formula $x = \frac{\epsilon y}{[1+(\epsilon-1)y]}$ when $\omega_s = 0$ with $\epsilon = \exp(b)$, which is the same result as that found by Priante et al. empirically [195]. Priante et al. showed that for the AlGaAs—GaAs interface when the Al flux is switched off, then the system is simplified, and

$$\frac{dy_{\text{Al}}}{d\zeta} = -\frac{g}{h}\frac{\epsilon y_{\text{Al}}}{1 + (\epsilon - 1)\, y_{\text{Al}}} \tag{1.27}$$

with ϵ the ratio between reaction quotients of the $Al^l + Al^s \rightarrow AlAs^s$ and $Ga^l + As^l \rightarrow GaAs^s$ reactions. With boundary conditions $y(0) = y_0$, the solution of this leads to the relation

$$x\,(\zeta) \approx \frac{W\left[\epsilon y_0 \exp\left(\epsilon y_0 - \epsilon g \frac{\zeta}{h}\right)\right]}{1 + W\left[\epsilon y_0 \exp\left(\epsilon y_0 - \epsilon g \frac{\zeta}{h}\right)\right]} \tag{1.28}$$

where W is the principal branch of the Lambert function, which is defined by $W\,(z)\,e^{W(z)} = z$ [198]. In the model presented here, the dependence of interface sharpness on radius can be seen from the parameter g, which is given by [195]

$$g = 4\sqrt{3}\frac{\Omega_{\text{Ga}}}{a_0^2 R_{\text{NW}}}\frac{\sin^3\beta}{(1 - \cos\beta)^2(2 + \cos\beta)} \tag{1.29}$$

The second method considered by Glas et al. to find the relation between composition of the liquid catalyst and the solid is based on the classical nucleation theory [199]. It accounts for the fact that NWs grow far from equilibrium and that growth of each ML is mediated by the formation of a 2D nucleus at the solid-liquid interface. To find a relation for the solid composition, Reiss theory is used which states that the critical 2D nucleus is given by the saddle point of the surface describing the work of formation of the nucleus in the (composition,size) parameter space [199]. It is then assumed that a new ML adopts the composition of the critical nucleus.

The nucleation model uses the change in free energy ΔG of the system upon forming a nucleus of a specific size and composition from a liquid. The critical nucleus is given by $\partial\Delta G/\partial n_A = 0$, $\partial\Delta G/\partial n_B = 0$, and the size of the critical radius is determined by $\frac{\partial\Delta G}{\partial r} = 0$ [23]. Composition of the critical nucleus is given by the solution of

$$\Omega(1-2x) + k_B T \ln \frac{x}{1-x} = \mu_{Al}^{L} - \mu_{Al,As}^{AlAs} - (\mu_{Ga}^{L} - \mu_{Ga,As}^{GaAs}) \tag{1.30}$$

And the critical radius is given by

$$r_c = \frac{\alpha_1}{2\alpha_2} \frac{\omega_p^S \gamma_e}{\Delta\mu(x_c)} \tag{1.31}$$

The critical composition gives the required relation $y(x)$ needed to produce a composition profile of the heterostructure interface.

1.4.3 Muraki Model

A model which has attained a great deal of attention from work regarding thin film growth and interfaces of thin film heterostructures is the Muraki model [200]. This model looks at surface segregation during MBE, with the sharpness of heterointerfaces being of interest. The GaAsP-GaAs material system is considered here as an illustrative example for the Muraki model applied to a QW/QD.

In the model developed by Muraki et al. it is assumed a fraction R of P atoms on the topmost layer segregate to the next layer during growth. That is, $(1 - R)$ of the P atoms are incorporated into the bulk before the next ML is completed. The P composition in the nth layer is given by

$$x_n = x_0(1 - R^n) \qquad (1 \leq n \leq N;\ \text{well}) \tag{1.32}$$

$$x_n = x_0(1 - R^N)R^{n-N} \qquad (n > N;\ \text{barrier}) \tag{1.33}$$

where x_0 is the nominal P composition and N the well/dot width in MLs. R is the segregation probability and is described by

$$R = \exp\left(\frac{-d}{\lambda}\right) \tag{1.34}$$

where λ is the segregation length and d is half the lattice constant of GaAs.

1.4.4 Empirical Sigmoidal Models

While models based on the physical process of material growths are likely to be more accurate and better represent the system of interest, these models can make analysis of experimental samples difficult when the models are relatively complex with many parameters. It can sometimes be useful to use empirical models that simply fit the shape of the data profile so that quantities of interest can be more easily extracted.

For heterostructure interfaces, it is common to use sigmoidal type fits to interfaces [201]. Depending on what information is required from the data, different functions are available to use. An example of such an equation used to describe the profile of (Ga,In)(N,As) QWs is [201]

$$x_n = \frac{x_0^{(l)}}{1 + e^{-\frac{(n - N_w/2)}{L^{(l)}}}} \tag{1.35}$$

where N_w is the width of QW in MLs, and L is a fitting parameter the defines the range around the inflection point $n = \pm N_w/2$ where the concentration changes from $x_0/(1 + e^{1/2})$ to $x_0/(1 + e^{-1/2})$. The width of the interface W is often estimated by using 10% to 90% variation, which is defined as the length over which the concentration changes from 10% to 90% of its plateau value. Width of the interface is then given by $W \approx 4.4L$.

Another option is the Boltzmann sigmoidal fit, which is given by

$$y = \frac{A_1 - A_2}{1 + e^{(x - x_0)/dx}} + A_2 \tag{1.36}$$

A_1 and A_2 are the two values y varies between either side of the interface, x_0 is the centre of the interface and dx is related to the gradient of the interface. A straight line gradient can be obtained for the centre point using

$$y' = \frac{A_2 - A_1}{4dx} \tag{1.37}$$

which can be useful when it comes to analysis of interface sharpness, since a sharper interface will have a steeper gradient.

1.4.5 Measuring Heterostructure Features Using STEM

While it has been established that heterostructures in NWs have a range of potentially useful features/applications, in order to test any samples grown a method is required to identify heterostructure features and check the quality of samples grown. There are a few options available, such as secondary ion mass spectrometry (SIMS) [202, 203], however this section will focus on scanning transmission electron microscope (STEM), and looks at its use in obtaining composition profiles of heterostructures.

In NW heterostructures, the main reason for contrast in high angle annular dark field (HAADF) STEM images is chemical variation, where image intensity changes with composition. The compositional sensitivity of this technique is usually seen at large scattering angles in STEM imaging, and more details about this can be seen in Sect. 2.1.5. An example of this compositional sensitivity was seen in the study from Grillo et al. [204] where GaAs QWs were studied in InGaAs thin films. They pointed out that for thin specimens the surface relaxation affects strain fields in the sample, with this phenomenon known to influence TEM contrast [205, 206]. Another effect that contributes to contrast is dechannelling, which is when interband scattering between Bloch waves induced by local strain causes a reduction of the 1s contribution to the electron wavefunction [204]. At the high scattering angles in HAADF images, dechannelling causes a reduction in image intensity [187].

A common feature of a heterostructure with a change in composition will be mismatch of the crystal lattices, which cubic materials respond to by tetragonal distortions [204]. In many cases this will have strain associated with it. For thin samples there will also be lattice bending which results from surface relaxation [187]. This was first identified by Grillo et al. [204] and has also been reported by Beyers et al. [187]. Grillo noticed that small tilts off zone axis conditions reduces HAADF intensity, and saw dips in intensity on either side of a GaAs QW inserted into an InGaAs thin film growth. This causes dark bands at the interface in lower magnification images [187]. Intensity minima vary in depth and position depending on experimental conditions, with tilting the sample causing intensity minimum to shift from the left QW interface to the right. At large tilt angles it was seen that one of the intensity minima can disappear while the other is greatly enhanced.

The observed behaviour is explained by a mix of the lattice bending causing dechannelling, and tilting. In on-axis conditions, HAADF intensity is increased from channelling, and lattice plane bending results in lower excitation of 1s and causes a reduction in intensity at and close to the interface. Tilting the sample causes a reduction in the excitation of the columnar states and intensity is reduced [204]. A small tilt off axis can partially compensate for the dechannelling at one interface while enhancing the effect at the other.

In experimental samples, interpretation of images can be complicated further by superposition of other effects on top of this. At heterostructure interfaces, things like compositional segregation can also alter intensity. Some broadening of intensity change can come from the fact that the probe has a finite size because of aberrations and source size [187]. With all these effects possible, care must be taken interpreting images, where small mistilts can drastically alter image intensity. A reduction in intensity could be misinterpreted as a change in composition.

1.5 Thesis Outline

The work presented in this thesis takes advantage of the resolving power of aberration-corrected electron microscopy to study the defective structure of self-catalysed GaAsP NWs that forms during the droplet consumption stage with sub-optimal growth conditions. Defects present in a NW can diminish optoelectronic properties, and so defect stability is investigated to see if defects can be removed from the NW system.

Chapter 2 looks at techniques used to produce the data presented in the results chapters. A large section of this chapter looks at the types of electron microscopy used since this is what enables structural features to be imaged and analysed. This chapter also looks into the types of sample preparation required to perform the electron microscopy, along with types of spectroscopy used.

Chapter 3 looks into the types of defects observed inside NWs using atomic resolution STEM. A procedure to identify Burgers vector is used to analyse a wide variety of defect structures. The origin of these defects is investigated via in-situ heating experiments inside a TEM, and the effect of their presence on optoelectronic properties is tested using cathodoluminescence (CL).

In Chap. 4, the stability of some of the defects observed in Chap. 3 is tested using in-situ heating experiments inside a TEM. Motion (or in some cases lack of motion) is tracked over time, and by measuring velocities, an estimate activation energy value for the most commonly observed defect type is found.

Finally, in Chap. 5, interfaces of a GaAsP NW with GaAs QDs are examined. Different models that describe heterostructure interfaces are tested to see how well each fits experimental data. The relation between interface sharpness and radius is tested along with the size of QDs for a range of NW radii.

References

1. Sanchez AM et al (2017) Nonradiative step facets in semiconductor nanowires. Nano Lett 17:2454–2459
2. Zhang A, Zheng G, Lieber C (2016) Nanowires building blocks for nanoscience and nanotechnology. Springer
3. Arbiol J, Xiong Q (2015) Semiconductor nanowires. Woodhead Publishing
4. Fukata N, Rurali R (2021) Fundamental properties of semiconductor nanowires. Springer
5. McIntyre P, Fontcuberta i Morral A (2020) Semiconductor nanowires: to grow or not to grow? Mater Today Nano 9:100058
6. Güniat L, Caroff P, Fontcuberta i Morral A (2019) Vapor phase growth of semiconductor nanowires: key developments and open questions. Chem Rev 119:8958–8971
7. Wong-Leung J et al (2020) Engineering III-V semiconductor nanowires for device applications. Adv Mater 32:1904359
8. Jia C, Lin Z, Huang Y, Duan X (2019) Nanowire electronics: from nanoscale to macroscale. Chem Rev 119:9074–9135
9. Royo M, Luca MD, Rurali R, Zardo I (2017) A review on III-V core-multishell nanowires: growth, properties, and applications. J Phys D: Appl Phys 50:143001
10. Wang S, Shan Z, Huang H (2017) The mechanical properties of nanowires. Adv Sci 4:1600332
11. Zhang Y, Wu J, Aagesen M, Liu H (2015) III-V nanowires and nanowire optoelectronic devices. J Phys D: Appl Phys 48:463001
12. Johansson J, Dick KA (2011) Recent advances in semiconductor nanowire heterostructures. CrystEngComm 13:7175–7184
13. Johansson J et al (2006) Structural properties of ⟨111⟩ B-oriented III-V nanowires. Nat Mater 5:574–580
14. Chang S, Lee GJ, Song YM (2020) Recent advances in vertically aligned nanowires for photonics applications. Micromachines 11
15. Wang Y et al (2013) Silicon nanowires for biosensing, energy storage, and conversion. Adv Mater 25:5177–5195
16. Murray CB, Kagan CR, Bawendi MG (1995) Self-organization of CdSe nanocrystallites into three-dimensional quantum dot superlattices. Science 270:1335–1338

17. Treuting R, Arnold S (1957) Orientation habits of metal whiskers. Acta Metall 5:598
18. Greiner ES, Gutowski JA, Ellis WC (1961) Preparation of silicon ribbons. J Appl Phys 32:2489–2490
19. Wagner RS, Ellis WC (1964) Vapor-liquid-solid mechanism of single crystal growth. Appl Phys Lett 4:89–90
20. Hu J, Ouyang M, Yang P, Lieber C (1999) Controlled growth and electrical properties of heterojunctions of carbon nanotubes and silicon nanowires. Nature 399:48–51
21. Duan X, Lieber CM (2000) Laser-assisted catalytic growth of single crystal GaN nanowires. J Am Chem Soc 122:188–189
22. Wu Y, Yang P (2000) Germanium nanowire growth via simple vapor transport. Chem Mater 12:605–607
23. Glas F (2017) Comparison of modeling strategies for the growth of heterostructures in III-V nanowires. Cryst Growth Des 17:4785–4794
24. Dubrovskii VG, Sibirev NV (2016) Factors influencing the interfacial abruptness in axial III-V nanowire heterostructures. Cryst Growth Des 16:2019–2023
25. Dubrovskii VG, Hijazi H (2020) Effect of arsenic depletion on the silicon doping of vapor-liquid-solid GaAs nanowires. Phys Status Solidi (RRL)-Rapid Res Lett 14:2000129
26. Morales AM, Lieber CM (1998) A laser ablation method for the synthesis of crystalline semiconductor nanowires. Science 279:208–211
27. Noborisaka J, Motohisa J, Fukui T (2005) Catalyst-free growth of GaAs nanowires by selective-area metalorganic vapor-phase epitaxy. Appl Phys Lett 86:213102
28. Jensen LE et al (2004) Role of surface diffusion in chemical beam epitaxy of InAs nanowires. Nano Lett 4:1961–1964
29. Zhang Y et al (2013) Self-catalyzed GaAsP nanowires grown on silicon substrates by solid-source molecular beam epitaxy. Nano Lett 13:3897–3902
30. Holmes JD, Johnston KP, Doty RC, Korgel BA (2000) Control of thickness and orientation of solution-grown silicon nanowires. Science 287:1471–1473
31. Law M, Goldberger J, Yang P (2004) Semiconductor nanowires and nanotubes. Annu Rev Mater Res 34:83–122
32. Heitsch AT, Fanfair DD, Tuan H-Y, Korgel BA (2008) Solution-liquid-solid (SLS) growth of silicon nanowires. J Am Chem Soc 130:5436–5437
33. Lensch-Falk JL, Hemesath ER, Lopez FJ, Lauhon LJ (2007) Vapor-solid-solid synthesis of Ge nanowires from vapor-phase-deposited manganese germanide seeds. J Am Chem Soc 129:10670–10671
34. Zhang R, Lifshitz Y, Lee S (2003) Oxide-assisted growth of semiconducting nanowires. Adv Mater 15:635–640
35. Hobbs RG, Petkov N, Holmes JD (2012) Semiconductor nanowire fabrication by bottom-up and top-down paradigms. Chem Mater 24:1975–1991
36. Wang GT, Li Q, Wierer JJ, Koleske DD, Figiel JJ (2014) Top-down fabrication and characterization of axial and radial III-nitride nanowire LEDs. Phys Status Solidi (a) 211:748–751
37. Dick KA, Caroff P (2014) Metal-seeded growth of III-V semiconductor nanowires: towards gold-free synthesis. Nanoscale 6:3006–3021
38. Schmid H et al (2015) Template-assisted selective epitaxy of III-V nanoscale de- vices for co-planar heterogeneous integration with Si. Appl Phys Lett 106:233101
39. Knoedler M et al (2017) Observation of twin-free GaAs nanowire growth using template-assisted selective epitaxy. Cryst Growth Des 17:6297–6302
40. Mayer B et al (2018) Microcavity III-V lasers monolithically grown on silicon in quantum sensing and nano electronics and photonics XV. In: Razeghi M, Brown GJ, Lewis JS, Leo G (eds) vol 10540. SPIE, pp 167–172
41. Vukajlovic-Plestina J et al (2019) Fundamental aspects to localize self-catalyzed III-V nanowires on silicon. Nat Commun 10:869
42. Dick KA (2008) A review of nanowire growth promoted by alloys and non-alloying elements with emphasis on Au-assisted III-V nanowires. Prog Cryst Growth Charact Mater 54:138–173

43. Messing ME, Hillerich K, Johansson J, Deppert K, Dick KA (2009) The use of gold for fabrication of nanowire structures. Gold Bull 42:172–181
44. Colombo C, Spirkoska D, Frimmer M, Abstreiter G, Fontcuberta i Morral A (2008) Ga-assisted catalyst-free growth mechanism of GaAs nanowires by molecular beam epitaxy. Phys Rev B 77:155326
45. Kim W et al (2018) Bistability of contact angle and its role in achieving quantum-thin self-assisted GaAs nanowires. Nano Lett 18:49–57
46. Ross FM (2010) Controlling nanowire structures through real time growth studies. Rep Prog Phys 73:114501
47. Jacobsson D et al (2016) Interface dynamics and crystal phase switching in GaAs nanowires. Nature 531:317–322
48. Harmand J-C et al (2018) Atomic step flow on a nanofacet. Phys Rev Lett 121:166101
49. Koivusalo ES, Hakkarainen TV, Guina MD, Dubrovskii VG (2017) Sub-poissonian narrowing of length distributions realized in Ga-catalyzed GaAs nanowires. Nano Lett 17:5350–5355
50. Glas F, Dubrovskii VG (2017) Self-narrowing of size distributions of nanostructures by nucleation antibunching. Phys Rev Mater 1:036003
51. Dubrovskii V (2014) Nucleation theory and growth of nanostructures. Springer, Berlin
52. Dai X, Olivier A, Wilhelm C, Dayeh S, Soci C (2015) In: Arbiol J, Xiong Q (eds) Semiconductor nanowires. Woodhead Publishing, pp 71–124
53. Growth and characterization of defect free GaAs nanowires. J Cryst Growth 287. The 16th American conference on crystal growth and epitaxy the 12th biennial workshop on OMVPE, 504–508 (2006)
54. Wang J et al (2013) Reversible switching of InP nanowire growth direction by catalyst engineering. Nano Lett 13:3802–3806
55. Yuan X et al (2015) Tunable polarity in a III-V nanowire by droplet wetting and surface energy engineering. Adv Mater 27:6096–6103
56. Kelly A, Knowles K (2012) Crystallography and crystal defects. Wiley
57. Zhang Y et al (2015) Polarity-driven quasi-3-fold composition symmetry of self-catalyzed III-V-V ternary core-shell nanowires. Nano Lett 15:3128–3133
58. Hiruma K et al (1993) GaAs free-standing quantum-size wires. J Appl Phys 74:3162–3171
59. Glas F, Harmand J-C, Patriarche G (2007) Why does wurtzite form in nanowires of III-V zinc blende semiconductors? Phys Rev Lett 99:146101
60. Caroff P, Bolinsson J, Johansson J (2011) Crystal phases in III-V nanowires: from random toward engineered polytypism. IEEE J Sel Top Quantum Electron 17:829–846
61. Dick KA, Thelander C, Samuelson L, Caroff P (2010) Crystal phase engineering in single InAs nanowires. Nano Lett 10:3494–3499
62. Lehmann S, Jacobsson D, Deppert K, Dick K (2012) High crystal quality wurtzite-zinc blende heterostructures in metal-organic vapor phase epitaxy-grown GaAs nanowires. Nano Res 5:470–476
63. Panciera F et al (2020) Phase selection in self-catalyzed GaAs nanowires. Nano Lett 20:1669–1675
64. Hofmann S et al (2008) Ledge-flow-controlled catalyst interface dynamics during Si nanowire growth. Nat Mater 7:372–375
65. Dick K et al (2004) Synthesis of branched "nanotrees" by controlled seeding of multiple branching events. Nat Mater 3:380–384
66. Mohammad NS (2014) Understanding quantum confinement in nanowires: basics, applications and possible laws. J Phys: Condens Matter 26:423202
67. Gaponenko SV (1998) Optical properties of semiconductor nanocrystals. Cambridge University Press
68. Efros AL, Rosen M (2000) The electronic structure of semiconductor nanocrystals. Annu Rev Mater Sci 30:475–521
69. Yoffe AD (2002) Low-dimensional systems: quantum size effects and electronic properties of semiconductor microcrystallites (zero-dimensional systems) and some quasi-two-dimensional systems. Adv Phys 51:799–890

70. Yu H, Li J, Loomis RA, Wang L-W, Buhro WE (2003) Wires on water. Nat Mater 2:517–520
71. Zuo J, Spence J (2017) Advanced transmission electron microscopy imaging and diffraction in nanoscience. Springer
72. Kasap S, Capper P (2017) Springer handbook of electronic and photonic materials. Springer
73. Adachi S (2017) In: Kasap S, Capper P (eds) Springer handbook of electronic and photonic materials. Springer International Publishing, Cham
74. De A, Pryor CE (2010) Predicted band structures of III-V semiconductors in the wurtzite phase. Phys Rev B 81:155210
75. Assali S et al (2013) Direct band gap wurtzite gallium phosphide nanowires. Nano Lett 13:1559–1563
76. Assali S et al (2015) Exploring crystal phase switching in GaP nanowires. Nano Lett 15:8062–8069
77. Akopian N, Patriarche G, Liu L, Harmand J-C, Zwiller V (2010) Crystal phase quantum dots. Nano Lett 10:1198–1201
78. Vainorius N et al (2015) Confinement in thickness-controlled GaAs polytype nanodots. Nano Lett 15:2652–2656
79. Karg S et al (2013) Measurement of thermoelectric properties of single semiconductor nanowires. J Electron Mater 42:2409–2414
80. Li D et al (2003) Thermal conductivity of individual silicon nanowires. App Phys Lett 83:2934–2936
81. Harada T et al (2020) Decreasing of the thermal conductivity of Si nanopillar/SiGe composite films investigated by using a piezoelectric photothermal spectroscopy. Jpn J Appl Phys 59, SKKA08
82. Tomioka K, Motohisa J, Hara S, Hiruma K, Fukui T (2010) GaAs/AlGaAs core multishell nanowire-based light-emitting diodes on Si. Nano Lett 10:1639–1644
83. Cao L et al (2009) Engineering light absorption in semiconductor nanowire devices. Nat Mater 8:643–647
84. Krogstrup P et al (2013) Single-nanowire solar cells beyond the Shockley-Queisser limit. Nat Photonics 7:306–310
85. Grzela G et al (2012) Nanowire antenna emission. Nano Lett 12:5481–5486
86. Kim H-M et al (2004) High-brightness light emitting diodes using dislocation- free indium gallium nitride/gallium nitride multiquantum-well nanorod arrays. Nano Lett 4:1059–1062
87. Signorello G et al (2014) Inducing a direct-to-pseudodirect bandgap transition in wurtzite GaAs nanowires with uniaxial stress. Nat Commun 5:3655
88. Balaghi L et al (2019) Widely tunable GaAs bandgap via strain engineering in core/shell nanowires with large lattice mismatch. Nat Commun 10:2793
89. Assali S et al (2019) Strain engineering in Ge/GeSn core/shell nanowires. Appl Phys Lett 115:113102
90. Nasr Esfahani M, Alaca BE (2019) A review on size-dependent mechanical properties of nanowires. Adv Eng Mater 21:1900192
91. Wu JY, Nagao S, He JY, Zhang ZL (2011) Role of five-fold twin boundary on the enhanced mechanical properties of fcc Fe nanowires. Nano Lett 11:5264–5273
92. Chen B et al (2013) Strengthening brittle semiconductor nanowires through stacking faults: insights from in situ mechanical testing. Nano Lett 13:4369–4373
93. Haraguchi K, Katsuyama T, Hiruma K, Ogawa K (1992) GaAs p-n junction formed in quantum wire crystals. Appl Phys Lett 60:745–747
94. Kelly MJ, Palmer RE, Lauhon LJ, Gudiksen MS, Lieber CM (2004) Semiconductor nanowire heterostructures. Philos Trans R Soc London Ser A: Math, Phys Eng Sci 362:1247–1260
95. Thelander C et al (2006) Nanowire-based one-dimensional electronics. Mater Today 9:28–35
96. Björk MT, Knoch J, Schmid H, Riel H, Riess W (2008) Silicon nanowire tunneling field-effect transistors. Appl Phys Lett 92:193504
97. Lu W, Lieber C (2007) Nanoelectronics from the bottom up. Nat Mater 6:841–850
98. Colinge J-P et al (2010) Nanowire transistors without junctions. Nat Nanotechnol 5:225–229

99. Thelander C et al (2003) Single-electron transistors in heterostructure nanowires. Appl Phys Lett 83:2052–2054
100. Hocevar M et al (2012) Growth and optical properties of axial hybrid III-V/silicon nanowires. Nat Commun 3:1266
101. Ren D et al (2018) Single-mode near-infrared lasing in a GaAsSb-based nanowire superlattice at room temperature. Nano Lett 18:2304–2310
102. Gmachl C, Capasso F, Sivco DL, Cho AY (2001) Recent progress in quantum cascade lasers and applications. Rep Prog Phys 64:1533–1601
103. Tribu A et al (2008) A high-temperature single-photon source from nanowire quantum dots. Nano Lett 8:4326–4329
104. Sköld N et al (2009) Microphotoluminescence studies of tunable wurtzite InAs0.85P0.15 quantum dots embedded in wurtzite InP nanowires. Phys Rev B 80:041312
105. Minot ED et al (2007) Single quantum dot nanowire LEDs. Nano Lett 7:367–371
106. Wilson DJ et al (2019) Integrated gallium phosphide nonlinear photonics. Nat Photonics
107. Yang Z et al (2019) Single-nanowire spectrometers. Science 365:1017–1020
108. Pettersson H et al (2006) Infrared photodetectors in heterostructure nanowires. Nano Lett 6:229–232
109. Ren D, Ahtapodov L, van Helvoort ATJ, Weman H, Fimland B (2019) Epitaxially grown III-arsenide-antimonide nanowires for optoelectronic appli- cations. Nanotechnology 30:294001
110. Zhang H et al (2018) Quantized Majorana conductance. Nature 556:74–79
111. Doh Y-J et al (2005) Tunable supercurrent through semiconductor nanowires. Science 309:272–275
112. Gazibegovic S et al (2017) Epitaxy of advanced nanowire quantum devices. Nature 548:434–438
113. Vaitiek ėnas S et al (2018) Selective-area-grown semiconductor-superconductor hybrids: a basis for topological networks. Phys Rev Lett 121:147701
114. Thelander C, Nilsson HA, Jensen LE, Samuelson L (2005) Nanowire single- electron memory. Nano Lett 5:635–638
115. Nilsson HA, Thelander C, Fröberg LE, Wagner JB, Samuelson L (2006) Nanowire-based multiple quantum dot memory. Appl Phys Lett 89:163101
116. Patolsky F, Zheng G, Lieber CM (2006) Nanowire sensors for medicine and the life sciences. Nanomedicine 1:51–65
117. Hayden O, Agarwal R, Lu W (2008) Semiconductor nanowire devices. Nano Today 3:12–22
118. Kim W, Ng JK, Kunitake ME, Conklin BR, Yang P (2007) Interfacing silicon nanowires with mammalian cells. J Am Chem Soc 129:7228–7229
119. Lee C-H et al (2011) Flexible inorganic nanostructure light-emitting diodes fabricated on graphene films. Adv Mater 23:4614–4619
120. Dattoli EN et al (2007) Fully transparent thin-film transistor devices based on SnO_2 nanowires. Nano Lett 7:2463–2469
121. Cohen D, McKernan S, Carter CB (1999) Characterization of the absolute crystal polarity across twin boundaries in gallium phosphide using convergent-beam electron diffraction. Microsc Microanal 5:173–186
122. Karlsson LS et al (2007) Understanding the 3D structure of GaAs ⟨111⟩ B nanowires. Nanotechnology 18:485717
123. Hirth J, Lothe J (1982) Theory of dislocations. Krieger Publishing Company
124. Qian X, Kawai M, Goto H, Li J (2015) Effect of twin boundaries and structural polytypes on electron transport in GaAs. Comput Mater Sci 108:258–263
125. Chen T et al (1992) Study of twins in GaAs, GaP and InAs crystals. J Cryst Growth 118:109–116
126. Dasilva YAR, Kozak R, Erni R, Rossell MD (2017) Structural defects in cubic semiconductors characterized by aberration-corrected scanning transmission electron microscopy. Ultramicroscopy 176:11–22
127. De la Mata M et al (2012) Polarity assignment in ZnTe, GaAs, ZnO, and GaN-AlN nanowires from direct dumbbell analysis. Nano Lett 12:2579–2586

128. Peters JJ et al (2015) Artefacts in geometric phase analysis of compound materials. Ultramicroscopy 157:91–97
129. Pond RC, Vlachavas DS, Christian JW (1983) Bicrystallography. In: Proceedings of the royal society of London. A. mathematical and physical sciences, vol 386, pp 95–143
130. Smallman R, Ngan A (2014) Modern physical metallurgy. Butterworth-Heinemann
131. Hull D, Bacon D (2011) Introduction to dislocations. Elsevier
132. Khanikar P, Kumar A, Subramaniam A (2011) Image forces on edge dislocations: a revisit of the fundamental concept with special regard to nanocrystals. Philos Mag 91:730–750
133. Ye W, Ougazzaden A, Cherkaoui M (2013) Analytical formulations of image forces on dislocations with surface stress in nanowires and nanorods. Int J Solids Struct 50:4341–4348
134. Martínez E, Marian J, Arsenlis A, Victoria M, Perlado J (2008) Atomistically informed dislocation dynamics in fcc crystals. J Mech Phys Solids 56:869–895
135. Smith DJ (2020) Atomic-resolution structure imaging of defects and interfaces in compound semiconductors. Prog Cryst Growth Charact Mater 66:100498
136. Zamani RR, Arbiol J (2019) Understanding semiconductor nanostructures via advanced electron microscopy and spectroscopy. Nanotechnology 30:262001
137. Glas F (2006) Critical dimensions for the plastic relaxation of strained axial heterostructures in free-standing nanowires. Phys Rev B 74:121302
138. Madelung O (2004) Semiconductors: data handbook. Springer
139. Denton AR, Ashcroft NW (1991) Vegard's law. Phys Rev A 43:3161–3164
140. Brown E, Sheng C, Shimamura K, Shimojo F, Nakano A (2015) Enhanced charge recombination due to surfaces and twin defects in GaAs nanostructures. J Appl Phys 117:054307
141. Takeuchi S, Suzuki K (1999) Stacking fault energies of tetrahedrally coordinated crystals. Phys Status Solidi (a) 171:99–103
142. Yamaguchi M, Paek J-H, Amano H (2012) Probability of twin formation on self-catalyzed GaAs nanowires on Si substrate. Nanoscale Res Lett 7:558
143. Zamani RR, Gorji Ghalamestani S, Niu J, Sköld N, Dick KA (2017) Polarity and growth directions in Sn-seeded GaSb nanowires. Nanoscale 9:3159–3168
144. Cayron C et al (2009) Odd electron diffraction patterns in silicon nanowires and silicon thin films explained by microtwins and nanotwins. J Appl Crystallogr 42:242–252
145. Jeon N, Dayeh SA, Lauhon LJ (2013) Origin of polytype formation in VLS-grown Ge nanowires through defect generation and nanowire kinking. Nano Lett 13:3947–3952
146. Den Hertog MI et al (2011) Hidden defects in silicon nanowires. Nanotechnology 23:025701
147. Zamani M et al (2018) Optimizing the yield of A-polar GaAs nanowires to achieve defect-free zinc blende structure and enhanced optical functionality. Nanoscale 10:17080–17091
148. Xu L et al (2008) Structure and migration of (112) step on (111) twin boundaries in nanocrystalline copper. J Appl Phys 104:113717
149. Lu N, Du K, Lu L, Ye HQ (2014) Motion of 1/3 $\langle 111 \rangle$ dislocations on $\Sigma 3$ 112 twin boundaries in nanotwinned copper. J Appl Phys 115:024310
150. Marquis EA, Hamilton JC, Medlin DL, Léonard F (2004) Finite-size effects on the structure of grain boundaries. Phys Rev Lett 93:156101
151. Pénisson JM, Dahmen U, Mills MJ (1991) HREM study of a $\Sigma = 3112$ twin boundary in aluminium. Philos Mag Lett 64:277–283
152. Tanaka K, Kohyama M (2002) Experimental and theoretical study of $Sigma = 3$ incoherent twin boundary in β -SiC. Philos Mag A 82:215–229
153. Sawada H, Ichinose H, Kohyama M (2006) Gap states due to stretched bonds at the (112) $\Sigma 3$ boundary in diamond. J Phys: Condens Matter 19:026223
154. Paxton AT, Sutton AP (1988) A simple theoretical approach to grain boundaries in silicon. J Phys C: Solid State Phys 21:L481–L488
155. Kohyama M (2002) Computational studies of grain boundaries in covalent materials. Model Simul Mater Sci Eng 10:R31–R59
156. Sakaguchi N, Ichinose H, Watanabe S (2007) Atomic structure of faceted $\sum 3$ CSL grain boundary in silicon: HRTEM and ab-initio calculation. Mater Trans 48:2585–2589

157. Feng CB, Nie JL, Zu XT, Al-Jassim MM, Yan Y (2009) Structure and effects of vacancies in Σ3 (112) grain boundaries in Si. J Appl Phys 106:113506
158. Cohen D, Carter CB (2003) $\Sigma = 3$, 112 lateral twin boundaries in GaP. Interface Sci 11:391–401
159. Feng C et al (2012) Possible effects of oxygen in Te-rich Σ3 (112) grain boundaries in CdTe. Solid State Commun 152:1744–1747
160. Liu C-Y et al (2016) Self-passivation rule and structure of CdTe Σ3 (112) grain boundaries. Phys Rev B 93:205426
161. Lazebnykh VY, Mysovsky AS (2015) Ab initio and atomistic simulation of local structure and defect segregation on the tilt grain boundaries in silicon. J Appl Phys 118:135704
162. Hohenberg P, Kohn W (1964) Inhomogeneous electron gas. Phys Rev 136:B864–B871
163. Kohn W, Sham LJ (1965) Self-consistent equations including exchange and correlation effects. Phys Rev 140:A1133–A1138
164. Mauger A et al (1987) Theoretical study of the electronic structure of the incoherent 211 $\sum = 3$ grain boundary in Ge by the recursion approach. Phys Rev B 35:1267–1272
165. Bietsch A, Michel B (2002) Size and grain-boundary effects of a gold nanowire measured by conducting atomic force microscopy. Appl Phys Lett 80:3346–3348
166. Wu W, Brongersma SH, Van Hove M, Maex K (2004) Influence of surface and grain-boundary scattering on the resistivity of copper in reduced dimensions. Appl Phys Lett 84:2838–2840
167. Hemesath ER, Schreiber DK, Kisielowski CF, Petford-Long AK, Lauhon LJ (2012) Atomic structural analysis of nanowire defects and polytypes enabled through cross-sectional lattice imaging. Small 8:1717–1724
168. Hemesath ER et al (2012) Catalyst incorporation at defects during nanowire growth. Nano Lett 12:167–171
169. Allen JE et al (2008) High-resolution detection of Au catalyst atoms in Si nanowires. Nat Nanotechnol 3:168–173
170. Park G-S et al (2012) Full surface embedding of gold clusters on silicon nanowires for efficient capture and photothermal therapy of circulating tumor cells. Nano Lett 12:1638–1642
171. Webb JL et al (2017) Imaging atomic scale dynamics on III-V nanowire surfaces during electrical operation. Sci Rep 7:12790
172. Sumino K (1994) Mechanical behaviour of semiconductors. In: Handbook on semi- conductors. Elsevier Science, Amsterdam, pp 73–182
173. Mittemeijer E (2011) Fundamentals of materials science. Springer
174. Maeda K, Yamashita Y (1993) Dislocation motion in strained thin films. Are Kinks colliding with each other? Phys Status Solidi (a) 138:523–532
175. Mason W (1966) Physical acoustics principles and methods. Academic Press
176. Yonenaga I, Sumino K (1987) Effects of In impurity on the dynamic behavior of dislocations in GaAs. J Appl Phys 62:1212–1219
177. Yonenaga I, Sumino K (1989) Impurity effects on the generation, velocity, and immobilization of dislocations in GaAs. J Appl Phys 65:85–92
178. Chou Y-C, Panciera F, Reuter MC, Stach EA, Ross FM (2016) Nanowire growth kinetics in aberration corrected environmental transmission electron microscopy. Chem Commun 52:5686–5689
179. Yonenaga I, Sumino K (1993) Dynamic activity of dislocations in gallium phosphide. J Appl Phys 73:1681–1685
180. Chin AK, Keramidas VG, Johnston WD, Mahajan S, Roccasecca DD (1980) Evaluation of defects and degradation in GaAs-GaAlAs wafers using transmission cathodoluminescence. J Appl Phys 51:978–983
181. Maeda K, Yamashita Y, Maeda N, Takeuchi S (1990) Radiation enhanced dislocation glide and rapid degradation. MRS Proc 184:69
182. Lavagne S, Levade C, Vanderschaeve G (2006) Transmission electron microscopy in situ investigation of dislocation behaviour in semiconductors and the influence of electronic excitation. Philos Mag 86:4923–4940

183. Hyun JK, Zhang S, Lauhon LJ (2013) Nanowire heterostructures. Annu Rev Mater Res 43:451–479
184. Parkinson P et al (2009) Carrier lifetime and mobility enhancement in nearly defect-free core-shell nanowires measured using time-resolved terahertz spectroscopy. Nano Lett 9:3349–3353
185. Jiang N et al (2013) Enhanced minority carrier lifetimes in GaAs/AlGaAs coreshell nanowires through shell growth optimization. Nano Lett 13:5135–5140
186. Gudiksen MS, Lauhon LJ, Wang J, Smith DC, Lieber CM (2002) Growth of nanowire superlattice structures for nanoscale photonics and electronics. Nature 415:617–620
187. Beyer A et al (2017) Influence of surface relaxation of strained layers on atomic resolution ADF imaging. Ultramicroscopy 181:8–16
188. Madras P, Dailey E, Drucker J (2009) Kinetically induced kinking of vapor-liquid-solid grown epitaxial Si nanowires. Nano Lett 9:3826–3830
189. Jabeen F, Patriarche G, Glas F, Harmand J-C (2011) GaP/GaAs1-xPx nanowires fabricated with modulated fluxes: a step towards the realization of superlattices in a single nanowire. J Cryst Growth 323. Proceedings of the 16th international conference on molecular beam epitaxy (ICMBE), pp 293–296
190. Björk MT et al (2002) One-dimensional steeplechase for electrons realized. Nano Lett 2:87–89
191. Björk MT et al (2002) Nanowire resonant tunneling diodes. Appl Phys Lett 81:4458–4460
192. Clark TE et al (2008) Diameter dependent growth rate and interfacial abruptness in vapor-liquid-solid Si/Si1-xGex heterostructure nanowires. Nano Lett 8:1246–1252
193. Ramdani MR, Harmand JC, Glas F, Patriarche G, Travers L (2013) Arsenic pathways in self-catalyzed growth of GaAs nanowires. Cryst Growth Des 13:91–96
194. Priante G, Ambrosini S, Dubrovskii VG, Franciosi A, Rubini S (2013) Stopping and resuming at will the growth of GaAs nanowires. Cryst Growth Des 13:3976–3984
195. Priante G et al (2016) Sharpening the interfaces of axial heterostructures in self-catalyzed AlGaAs nanowires: experiment and theory. Nano Lett 16:1917–1924
196. Understanding the composition of ternary III-V nanowires and axial nanowire heterostructures in nucleation-limited regime. Mater Des 132:400–408 (2016)
197. Glas F, Ramdani MR, Patriarche G, Harmand J-C (2013) Predictive modeling of self-catalyzed III-V nanowire growth. Phys Rev B 88:195304
198. Corless RM, Gonnet GH, Hare DEG, Jeffrey DJ, Knuth DE (1996) On the Lambert W function. Adv Comput Math 5:329
199. Reiss H (1950) The kinetics of phase transitions in binary systems. J Chem Phys 18:840–848
200. Muraki K, Fukatsu S, Shiraki Y, Ito R (1993) Surface segregation of In atoms and its influence on the quantized levels in InGaAs/GaAs quantum wells. J Cryst Growth 127:546–549
201. Luna E, Ishikawa F, Batista P, Trampert A (2008) Indium distribution at the interfaces of (Ga, In)(N, As)/GaAs quantum wells. Appl Phys Lett 92:141913
202. Chia ACE, Boulanger JP, LaPierre RR (2013) Unlocking doping and compositional profiles of nanowire ensembles using SIMS. Nanotechnology 24:045701
203. Chia ACE et al (2015) Nanowire dopant measurement using secondary ion mass spectrometry. J Appl Phys 118:114306
204. Grillo V (2009) Quantitative evaluation of strain effects in STEM HAADF contrast. Microscopie 11:61–68
205. Treacy MMJ, Gibson JM (1986) The effects of elastic relaxation on transmission electron microscopy studies of thinned composition-modulated materials. J Vac Sci Technol B Microelectron Process Phenom 4:1458–1466
206. De Caro L, Giuffrida A, Carlino E, Tapfer L (1997) Effects of the elastic stress relaxation on the HRTEM image contrast of strained heterostructures. Acta Crystallogr Sect A 53:168–174

Chapter 2
Methods

2.1 Electron Microscopy

A TEM allows for the imaging of materials down to the nm scale, using accelerated electrons and a series of electromagnetic lenses to produce a magnified image of a specimen [1, 2]. By control of the lenses and using different detectors, a wide range of information can be obtained from a sample using a TEM. There are two main types of TEM operation, either transmission mode where a parallel electron beam is used to illuminate a sample to produce an image, or convergent beam mode where the electron beam is focused to a small point and scanned across a sample to build an image. A TEM has three main components, an illumination system, a stage and objective lens, and the image projection system. A schematic diagram of a TEM operating in both parallel and convergent mode is shown in Fig. 2.1 with the illumination system, stage and objective, and image forming system labelled by regions 1, 2, and 3 respectively. Lenses are represented by red disks and electron paths are represented by blue and purple lines.

In parallel mode an approximately flat wave front enters the sample where the electrons interact, scatter and diffract through the sample, and an exit wave emerges from the sample. The exit wave then goes through an objective lens and forms a DP at the back-focal plane. DPs are formed by Bragg reflections from planes of atoms in the sample and are associated with crystal planes [1]. An intermediate lens is used to either project the DP onto a screen/detector or to recombine them and project an image of the sample onto a screen/detector.

In convergent beam mode, the electron beam is condensed to a small point on the sample. Scan coils are used to adjust the position of the probe on the sample. The electrons interact and scatter from the sample, and these scattered electrons are detected at chosen scattering angles, with the lens system used to direct electrons to the desired detector. In STEM imaging, the image is built by assigning the number of detected electrons to a single pixel value before the beam moves to the next position on the sample.

J. A. Gott, *Defects in Self-Catalysed III-V Nanowires*, Springer Theses,
https://doi.org/10.1007/978-3-030-94062-1_2

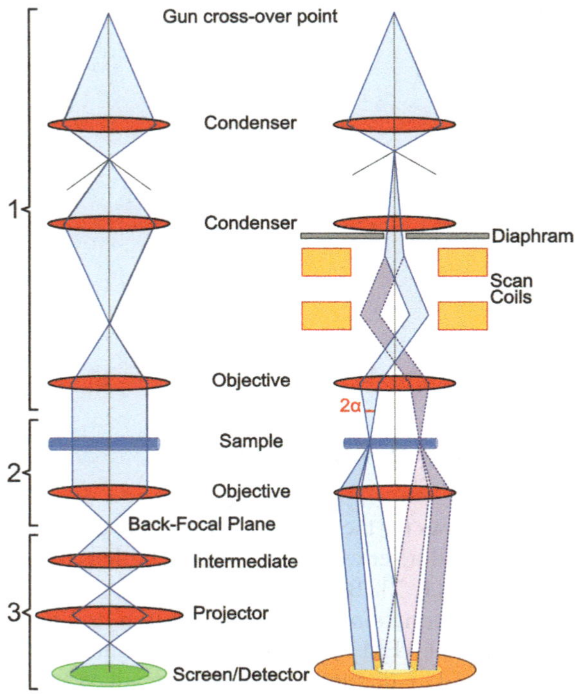

Fig. 2.1 Using lenses represented by red disks in a TEM to produce two types of electron beam illumination to image a sample. The path electrons take are shown by blue and purple lines. The optic axis is indicated by a dashed line through the centre of each TEM configuration. The left shows parallel beam illumination in imaging mode and the right shows convergent beam illumination. The right shows how scan coils can be used to adjust the position of the condensed probe on the sample and how the scattered electrons are directed to different detectors based on scattering angle

Taking a Fourier transform of either a TEM or STEM image gives a power spectrum that is equivalent to the DP of the same area of the sample [2].

Lenses in a TEM are equivalent to glass lenses used in visible light microscopy, and are used to control the path electrons take inside the TEM from fixed lens positions and varying the currents in the lens coils. Lenses are used to produce a magnified image of an object. One major disadvantage of these lenses is they cause spherical and chromatic aberrations, which can severely limit what can be achieved with a TEM, and limit resolution [2].

2.1.1 *Illumination System*

The first important component of a TEM is the electron source, with the job of providing high energy electrons to be sent through the TEM column to the sample. Typical accelerating voltages used vary roughly in the range 80–300 kV, and when voltages above roughly 100 kV are used, relativistic effects must be considered. The wavelength of an electron λ can be described by the relativistic relation [1]

$$\lambda = \frac{h}{\left[2m_0eV\left(1+\frac{eV}{2m_0c^2}\right)\right]^{\frac{1}{2}}} \tag{2.1}$$

where eV is the kinetic energy from accelerating voltage V, h is Planck's constant, m_0 is rest mass and c is speed of light in vacuum. In terms of accelerating voltage choice, a higher kV can potentially provide better image resolution because of a smaller wavelength, and can allow for imaging of thicker samples.

There are two main varieties of electron sources used in TEMs, and are either thermionic (which use heat to produce electrons) or field-emission (which use a large electric potential to produce electrons). Thermionic electron guns commonly use either tungsten or LaB_6 crystals. A smaller source gives better spacial coherence and is the reason why field-emission sources have better spatial coherence [1].

The electron source provides electrons which can be manipulated to provide two different modes of illumination to the sample, either a parallel beam for TEM imaging or selected-area diffraction, or a convergent beam for STEM imaging, convergent beam electron diffraction (CBED) and others such as energy dispersive x-ray spectroscopy (EDX). Both of these illumination modes are shown in Fig. 2.1. For parallel beam illumination, two condenser lenses are used to form the parallel beam. The first lens forms an image of the gun cross-over point and the second forms an image of the first lens crossover, and a third lens then forms a parallel beam.

For convergent beam mode, a condenser lens is used to control the probe size, and an aperture is used to control convergence angle. The diameter of the probe is influenced by a few things including source size (d_s), diffraction disc (d_d) and aberrations (d_g). Their contributions can be described by the equations [2]

$$d_s^2 = \frac{4I_{\text{probe}}}{\pi^2 B\alpha^2} \quad d_d^2 = \frac{(1.2\lambda)^2}{\alpha^2} \quad d_g^2 = A_n^2 C_{n,0}^2 \alpha^{2n} \tag{2.2}$$

where I_{probe} is probe current, B is brightness which is the current density per unit solid angle, α is the probe convergence angle and is labelled in Fig. 2.1, $C_{n,0}$ is the aberration coefficient, and A_n is a numerical constant that gives the diameter of the disc of least confusion. Aberrations and astigmatism can greatly impact the ability to form a small probe, and different aberrations become the most limiting contribution to probe size for different values of α. The position of the condensed beam can be changed using scanning coils. An example of the coils changing the electron beam path to direct the electron beam to two points on a sample is shown in Fig. 2.1.

2.1.2 Aberrations

For both parallel beam mode and convergent beam mode, aberrations caused by the lenses can limit achievable resolution. In convergent beam mode, aberrations affect probe formation, and in parallel beam mode they affect both the formation of the parallel beam and image formation with the objective lens. There are several types of aberrations that can alter electron behaviour in different ways [2]. Spherical aberrations occur when a lens behaves differently for off-axis electrons, and causes point-like objects to appear as a disk. Chromatic aberrations occur because the electrons do not all have the same energy and is the next most significant aberration that affects image resolution after spherical aberrations. The lens bends electrons of lower energy more strongly and causes a point-like object to appear as a disk. Electrons can have variation in energy because of fluctuations in the accelerating voltage of the gun or from interactions in the sample. Astigmatism is when the cross-over of off-axis rays is displaced along the optic axis, varying with azimuthal angle of the beam. Coma or comatic aberrations is an effect experienced by rays which are emitted from an object point off the optic axis, where off axis rays that do not travel through the centre of the lens are focussed at different points to those that travel through the centre of the lens.

A perfect lens take a plane wave and produces a perfect spherical wave that comes to a point focus. With aberrations, a phase change is introduced, and can be represented by the wave aberration function χ, defined as the phase difference between the perfect spherical wave and the actual wavefront for a given lens [2]. When condensing the beam to form a probe it ideally takes the shape of an Airy disc that is the Fourier transform of the aperture in the back focal plane of the lens. To include aberrations, the probe wave function $\psi_p(R)$ is calculated as the inverse Fourier transform of the aperture function $A(K)$ in the back focal plane. The aperture function includes a phase change $\exp(-i\chi(k))$ due to aberrations and an aperture that cuts off the wave at a particular angle. The probe wave function is given by [2]

$$\psi_p(R) = \int H(K) \exp(-i\chi(K)) \exp(-iR \cdot K)\, dK \tag{2.3}$$

where H is the aperture shape with value one inside the aperture and zero outside and K is a vector in the back focal plane of the probe forming lens. When lens aberrations tend to zero for rays close to the optic axis, then the aberration function can be described by the series [2]

$$\begin{aligned}
\chi(\theta, \phi) = {} & const + \theta\{C_{01a}\cos(\phi) + C_{01b}\sin(\phi)\} \\
& + \frac{\theta^2}{2}\{C_{10} + C_{12a}\cos(2\phi) + C_{12b}\sin(2\phi)\} \\
& + \frac{\theta^3}{3}\{C_{23a}\cos(3\phi) + C_{23b}\sin(3\phi) + C_{21a}\cos(\phi) + C_{21b}\sin(\phi)\} \\
& + \frac{\theta^4}{4}\{C_{30} + C_{34a}\cos(4\phi) + C_{34b}\sin(4\phi) + C_{32a}\cos(2\phi) + C_{32b}\sin(2\phi)\} \\
& + \ldots
\end{aligned} \tag{2.4}$$

Each term is of the form

$$\frac{\theta^{N+1}}{N+1}\{C_{NSa}\cos(S\phi) + C_{NSb}\sin(S\phi)\} \quad \text{or} \quad \frac{\theta^{N+1}}{N+1}C_{NS} \tag{2.5}$$

The coefficients C_{NS} are called the aberration coefficients. The C_{NSa} and C_{NSb} coefficients represent projections of the overall aberration C_{NS}. N describes radial order and S describes azimuthal symmetry. Each coefficient corresponds to a different kind of named aberrations, with for example C_{10} defocus, C_{21} coma and C_{30} spherical aberrations.

Aberration correctors can be used to compensate for the effect of some aberrations and improve the resolving power of the TEM. Correctors work by introducing negative aberrations to balance the aberrations introduced by the lenses.

2.1.3 Sample Interaction

When an electron encounters the sample, there are a few ways the electron can interact with the sample. Electron scattering will either result in no loss of energy (elastic scattering) where the electrons are usually coherent, or with loss of energy (inelastic scattering) where the electrons are usually incoherent [1]. Inelastic interactions can produce a variety of signals including Auger electrons, backscattered electrons, secondary electrons, x-rays, visible light, and electron-hole pairs. Each of these can provide different information about the sample. It is usually assumed in a TEM sample single scattering occurs. Some electrons will be scattered at an angle to the incident direction, and some will experience no deviation.

Elastic scattering mainly comes from interactions with atomic nuclei and usually described in terms of Rutherford scattering from an atom, which ignores screening by the electrons. Elastic scattering usually occurs at angles $<10°$ and becomes more incoherent at angles $>10°$ where scattering is mostly thermal diffuse. The cross section for elastic scattering is a function of Z and for unscreened Rutherford scattering the cross section is proportional to Z^2. At higher scattering angles, unscreened scattering is a good approximation, while at lower angles the scattering becomes more screened and less dependent on Z with intensity varying Z^{ζ} where $1.5 < \zeta < 2$. Thicker samples result in more scattering.

Inelastic scattering mainly involves electron-electron interactions and occurs at angles $<1°$. An unwanted side effect to inelastic scattering is that it can cause electron beam damage. Damage is caused by two main mechanisms, either knock-on damage where an atom or an ion is displaced from its normal site and can cause point defects, or ionisation damage which can cause chemical or structural changes. The beam interaction with the sample can also cause heating of the sample.

2.1.4 TEM

Contrast C in an image can be quantified by [1]

$$C = \frac{\Delta I}{I_1} \tag{2.6}$$

where ΔI is the difference in intensity of two adjacent regions of an image and I_1 is the lowest of the two intensities being compared. Contrast in TEM images comes from interference of the electron wave with itself and depends on sample thickness, density of the material, and on Z. For TEM imaging the sample is illuminated with a parallel electron beam. TEM imaging works using the idea that the objective lens takes electrons emerging from the sample, disperses them to make a DP in the back-focal plane and recombines them to form an image in the image plane, with these labelled in Fig. 2.1. A thin sample can be assumed to act as a weak phase object, which produces a spatially varying phase shift in the electron wave which contains information about the sample. A phase object modulates the incident wave by a specimen transmission function $\phi(R)$ [2]

$$\phi(R) = \exp(i\sigma V(R)) \tag{2.7}$$

where $\sigma = \frac{2\pi m e \lambda}{h^2}$ is an interaction constant, with m relativistic mass, and $V(R)$ is the sample potential. If the phase changes are small the transmission function can be simplified for a weak phase object

$$\phi(R) = \exp(i\sigma V(R)) \approx 1 + i\sigma V(R) \tag{2.8}$$

Without aberrations included, for an incident wave ψ_0 that has transmitted through the sample, the amplitude of the wave at the exit face of the sample ψ_{exit} is given by

$$\psi_{exit}(R) = \psi_0 \phi(R) \tag{2.9}$$

This also describes the wave function in the image plane. The diffraction amplitude in the back-focal plane $\tilde{\psi}_{exit}(K)$ is given by the Fourier transform of $\psi_{exit}(R)$

$$\tilde{\psi}_{exit}(K) = \mathcal{F}(\psi_0 \phi(R)) \tag{2.10}$$

The recorded image intensity is given by the product of the image wave function and its complex conjugate

$$I_{\text{image}(R)} = \psi_{\text{image}}(R)\psi^*_{\text{image}}(R) = |\psi_0|^2 \tag{2.11}$$

Effects of objective lens aberrations, partial coherence and finite size of the objective aperture can be included using a transfer function $A(K)$

$$A(K) = H(K)\exp(-i\chi(K)) \tag{2.12}$$

where $\chi(K)$ describes objective lens aberrations as seen in Sect. 2.1.2, and $H(K)$ describes effects of the objective aperture and partial coherence of the electron beam. The image amplitude is then given by the Fourier transform $\mathcal{F}\{\tilde{\psi}_{exit}(K)A(K)\}$, and image intensity $I(R)$ is then given by

$$I(R) = \psi_{\text{image}}(R)\psi^{*}_{\text{image}}(R) = 1 + 2\sigma V(R) \otimes \mathcal{F}\{H(K)\sin\chi(K)\} \tag{2.13}$$

The image can be seen using a fluorescent viewing screen or is usually recorded using a charge-coupled device (CCD). The recorded image contains only intensity of the wave resulting from the interference and phase information is lost. The diffraction pattern can also be projected onto the screen or CCD by adjusting the intermediate lens in the image projection system.

2.1.5 STEM

STEM imaging uses convergent beam illumination. An image is built by scanning the electron beam across the sample, the electrons interact with the sample as described in Sect. 2.1.3 and scattered electrons are detected from a chosen range of scattering angles. The probe is scanned across the sample and for each probe position, an image pixel intensity is given by the number of electrons detected by the detector. The magnification of the image is determined by the scan dimensions, and images are influenced by aberrations in the scanning probe. The resolution of STEM imaging is determined by the size of the probe formed by the lens which is limited by lens aberrations and aperture, as was seen in Sect. 2.1.3.

There are three commonly used detector positions for electron detection and image formation. The detector name and scattering angles used are bright-field (BF) which uses scattered angles in the range $\theta_1 < 10$ mrad, annular dark-field (ADF) which collects electrons scattered in the range $10 < \theta_2 < 50$ mrad, and HAADF which collects electrons scattered at angles $\theta_3 > 50$ mrad [1]. These detectors and angles are illustrated in Fig. 2.2. Contrast in STEM images is influenced by sample thickness and Z. A thicker sample or a sample with higher Z will scatter more electrons off axis compared to thinner or lower Z samples. In BF imaging, fewer electrons from thick/high Z regions reach the detector and so appear dark in BF images. The opposite to BF is true for ADF images. For detecting the scattered electrons to provide a value for a pixel in a STEM image, a scintillator-photomultiplier is normally used and requires a fast response from the detector.

BF imaging collects electrons scattered by a small angle, and so the electrons are mostly coherently scattered. Given that a sample, described by a transmission function $\phi(R)$ with R a position vector in the sample, is illuminated by a STEM probe with complex amplitude $\psi(R - R_p)$ located at R_p, then the wave function that exits the sample is given by the product of $\phi(R)$ and $\psi(R - R_p)$. The wave function

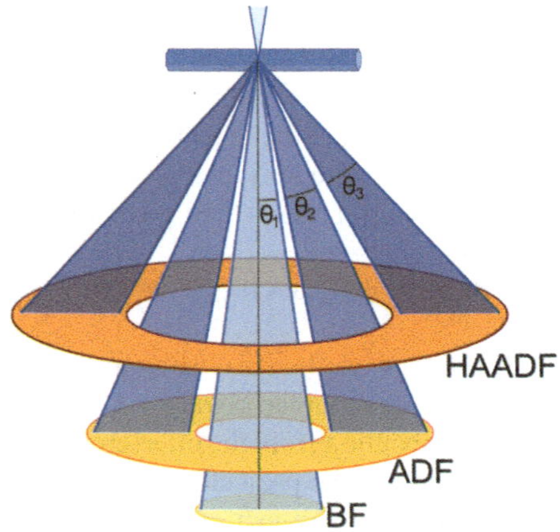

Fig. 2.2 Scattering of a condensed electron beam from a sample with scattering angles indicated for different detectors. For BF detectors, electrons scattered at angle $\theta_1 < 10\,\text{mrad}$ are collected, for ADF electrons scattered between $10\,\text{mrad} < \theta_2 < 50\,\text{mrad}$ are collected and for HAADF detectors electrons scattered at angle $\theta_3 > 50\,\text{mrad}$ are collected

observed at the detector is then given by the Fourier transform of the product of $\psi(R)$ and $\psi(R - R_p)$, and can be described by [2]

$$\psi_f(K_f, R_p) = \sum_g \phi_g A(K_f - g) \exp[-i(K_f - g) \cdot R_p] \tag{2.14}$$

where K_f is the transverse component of the wave scattered to the detector plane, A is the aperture function. This sum describes a series of diffracted disks forming a CBED pattern, with a simulated example of this shown in Fig. 2.3a with a CBED pattern of GaAs in a $\langle 110 \rangle$ direction. If the convergence angle is sufficiently large, the disks overlap and interfere and this leads to interference features that are sensitive to lens aberrations and probe position [3], with Fig. 2.3b showing a simulated pattern from the same area with a larger aperture used. Figure 2.3b shows an image with the right side being a simulated image and the left a schematic to make the disk overlap more clear. These simulated CBED images do not include thermal diffuse scattering (TDS). The yellow disk at the centre indicates the typical position of a BF detector. Variation of intensity from interference causes contrast in STEM images. For a specific spatial frequency g, the BF image intensity is [2]

$$I_{\text{BF}}(R_p) = 1 + 2|\sigma V_g| \cos(g \cdot R_p + \angle V_g) \sin \chi(g) \tag{2.15}$$

where $\angle V_g$ is the phase of the gth Fourier component of the sample potential. $\sin \chi(g)$ gives the strength at which each spatial frequency in the phase variation can contribute to the image, and is known as the phase contrast transfer function, and is identical to that found in TEM.

ADF imaging collects electrons from a slightly larger scattering angle compared to BF imaging. At the scattering angles used, an ADF detector will average over many interference features and this will suppress coherent phase information, and mostly incoherently scattered electrons reach the detector [2, 3]. A simulated CBED

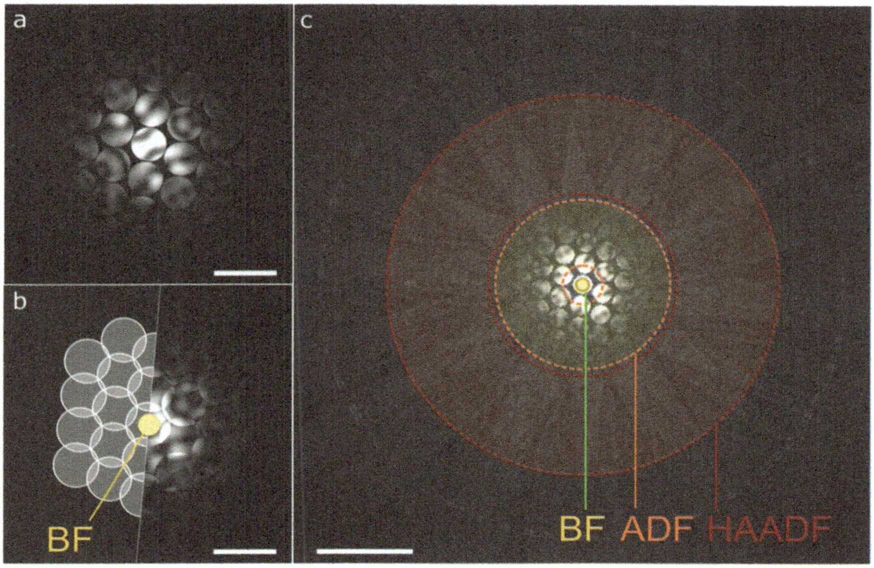

Fig. 2.3 CBED pattern simulations at 100 kV of GaAs in a ⟨110⟩ direction. **a** A pattern produced using beam angle 5.5 mrad where the disks do not overlap. **b** A pattern produced using convergent beam angle 7.0 mrad where the disks overlap and interfere. The image is split between a simulation on the right side and an illustration on the left side that shows disk positions and more clearly shows where the disks overlap. A yellow circle at the centre indicates where a typical BF detector is positioned. **c** CBED pattern produced using convergent beam angle 5.0 mrad with example positions of which electrons are detected by different detectors, with BF at the centre (yellow circle), ADF surrounding the centre (orange circle), and HAADF the furthest from the centre (red circle). Scale bars in (**a**) and (**b**) 20 mrad, in (**c**) 50 mrad

pattern is shown in Fig. 2.3c and shows some example positions of ADF and HAADF detectors, where it can be seen that the detector area includes many CBED disks that will overlap with a sufficiently large aperture. At the higher scattering angles collected by a HAADF detector the scattered intensity is strongly dependent on Z [1, 2]. The electron intensity detected depends on the intensity of the illuminating probe and the fraction that is scattered to the detector. The image can be described by a convolution of the intensity of the STEM probe and an object probe function $O(R)$ that represents the fraction of intensity that each atom is able to scatter to the detector. ADF intensity is given by [2]

$$I_{ADF(R_p)} = \left|\psi_p\left(R_p\right)\right|^2 \otimes O\left(R_p\right) \tag{2.16}$$

where ψ_p is STEM probe amplitude. States that contribute most to the intensity collected by the ADF detectors are the states most localised on the atomic column. When the electron probe is located over an atomic column in a crystal, the electrons become trapped by the attractive potential of the atoms. This is known as channelling, and an ADF image can be thought of as a map of the strength of channelling as a function of probe position.

At the scattering angles electrons are detected at in ADF imaging, thermal lattice vibrations strongly influence the obtained images [1, 2]. Thermal lattice vibrations reduce the strength of elastically scattered electrons and cause the intensity to become more diffuse. This comes from interactions of incident electrons that create phonons and causes a small energy loss with phonon scattered electrons scattered to angles in the range 5 to 15 mrad. TDS intensity can be larger than intensity of elastic scattering for HAADF detectors and so in any simulated image the effect of TDS must be included.

A common issue encountered in STEM imaging is sample drift. Since STEM records pixels sequentially, drift causes image distortions, which can become more prominent if longer pixel dwell times are used. A way around this is to use drift correction techniques. A relatively simple drift correction technique involves taking a stack of fast-scan images, aligning them using cross-correlation and then averaging the aligned stack of images. An option that takes into account scan distortions as well as stage drift effects is Smart Align software [4].

2.1.6 SEM

In a conventional scanning electron microscope (SEM), an electron probe is condensed and an objective lens is used to focus the electron beam onto the surface of a sample [5]. Accelerating voltages used can vary in the range from a few hundred V up to 30 kV. The electron probe is scanned over the sample surface by scan coils. The condensed beam penetrates into the sample and the interaction volume has the approximate shape of a teardrop. The interaction volume can extend from 100 nm to 5 μm depending on the energy of the incident electron beam and sample [5], and a diagram to illustrate this is shown in Fig. 2.4. This volume is much larger than the incident probe, and represents behaviour of a large number of electrons. This figure demonstrates that the interaction volume gets larger if higher accelerating voltages are used. The depth the electrons can reach $R(\mu\mathrm{m})$ is described by [5]

$$R = \frac{0.0276 A E_0^{1.67}}{\rho Z^{0.89}} \tag{2.17}$$

where A is atomic weight (g mol^{-1}), E_0 is the electron beam energy (keV), ρ is density (g cm^{-3}) and Z is atomic number. The interaction volume on the right side of the figure indicates a variety of signals that result from the electron beam interaction and roughly where they come from inside the sample.

The incident electron beam can be elastically or inelastically scattered by the sample, and the signals produced include: SEs, backscattered electrons, Auger electrons, characteristic x-rays, photons, and Bremsstrahlung X-rays. Elastically scat-

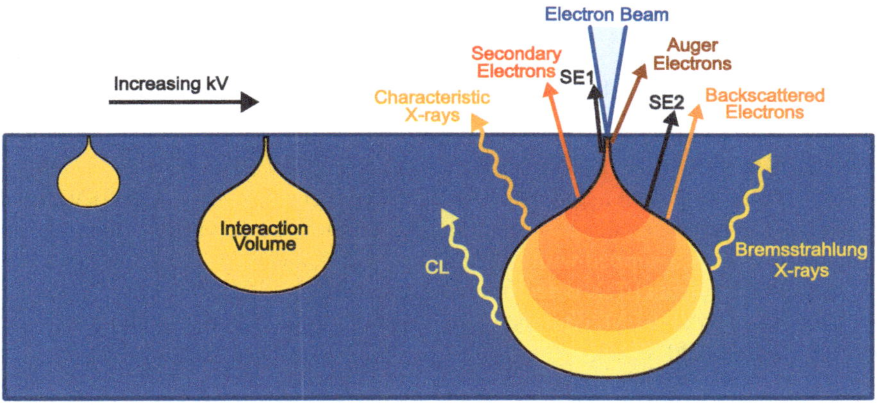

Fig. 2.4 Interaction of the electron beam with a sample inside a SEM. The total volume of sample that the incident electron beam interacts with is called the interaction volume and is roughly the shape of a teardrop. The interaction volume increases in size for larger accelerating voltages. The interaction of the beam with the sample produces a range of signals that can be detected and provide information about the sample

tered electrons are deflected through large angles and eventually leave the sample as backscattered electrons. Backscattered electrons make up a significant proportion of electron signals that come from the sample. When these electrons have left the sample, they can be captured by a detector and used to form a backscattered electron image. These images are sensitive to Z and posses atomic number contrast. SEs are electrons that are ejected from the sample after an inelastic interaction and are usually loosely bound outer shell electrons from atoms close to the sample surface, making this a surface sensitive technique. These can be collected to form a secondary electron image with contrast known as topographic contrast. Resolution of these images is limited by the probe size used. There are two commonly used detectors for two main types of SEs, SE1 and SE2 with both types of SEs shown in Fig. 2.4. An in-lens detector, which is positioned within the probe forming lens system, detects SE1 electrons that are generated near the top region of the interaction volume and provide direct information of the sample surface with high resolution. An E-T detector is usually positioned towards one side of the sample and detects both SE1 and SE2 electrons [5]. SE2 electrons are generated after multiple scattering inside the interaction volume and leave the sample at a greater distance away from the electron beam incident point, and carry lower resolution surface information. Images from an E-T detector provide better contrast at the cost of resolution.

2.1.7 Microscopes

Microscopy in this thesis has been performed using the following microscopes:

- JEOL 2100 TEM with a LaB_6 source operating at an accelerating voltage of 200 kV with a Gatan OneView CCD.
- JEOL ARM200F with a Schottkey field-emission gun operating at 200 kV, fitted with CEOS probe and image correctors. The Atomic Resolution analytical Microscope (ARM) is equipped with a Gatan Orius CCD, and ADF & BF detectors. An Oxford Instruments EDX detector is also attached.
- Zeiss Supra 55VP SEM, equipped with a Gatan MonoCL3 spectrometer.
- Zeiss Gemini 500 SEM.

In-situ heating of nanowire (NW)s to simulate NW growth was performed using the JEOL 2100 system.

All atomic resolution ADF imaging was performed using the ARM200F system. ADF images in this thesis use a probe with a convergence semiangle $\approx$ 25 mrad and a camera length of 8 cm which corresponds to scattering angles in the range 45 ± 5 to 180 ± 8mrad being detected. This scattering angle range corresponds to the commonly referred to high angle scattering region (50 to 200mrad) that forms HAADF images [1, 2, 6]. Unless otherwise stated, atomic resolution STEM images in this thesis have been filtered using a band-pass fast fourier transform (FFT) filter. To do this a FFT of the image is taken to obtain a power spectrum. The central area of the FFT is masked and removed up to the first spots that correspond to the crystal lattice. An inverse fast fourier transform (IFFT) of the masked FFT is taken to produce the filtered image. In this thesis the phrase "band-pass filtered" refers to this method of filtering.

2.2 Sample Preparation

2.2.1 Sample Growth

The NWs shown in Chaps. 3 and 4 were grown using solid-source MBE. GaAsP core NWs were grown using the parameters listed in the core column of Table 2.1. After core growth, the catalyst droplets were consumed by closing Ga flux and increasing the group V flux to 8×10^{-6} Torr for 30 min. More information about the growth procedure can be found in [7].

The NWs shown in Chap. 5 were grown using the method reported by Zhang et al. [8]. This technique first grows small GaAs stems using the conditions listed in the 'GaAs Stem' column of Table 2.2, and growth is switched to GaAsP using the conditions listed in the 'GaAsP Section' column of the same table. This growth results in a distribution of NW diameters. A series of GaAs segments were grown axially by turning off the P flux for periods of 10, 20, 30, 40, and 50 s with GaAsP

Table 2.1 List of growth parameters used to grow GaAsP NWs

Parameter	Core	Shell
Ga beam equivalent pressure	8.41×10^{-8} Torr	8.41×10^{-8} Torr
V/III flux ratio	44	50
P/(As+P) flux ratio	0.16	0.3
Substrate temperature	640 °C	400 °C
Growth time	1 h	1 h

Table 2.2 List of growth parameters used to grow GaAsP NWs with GaAs QDs with a range of diameters

Parameter	GaAs Stem	GaAsP Section
Ga beam equivalent pressure	1.12×10^{-7} Torr	1.12×10^{-7} Torr
V/III flux ratio	60	50
P/(As+P) flux ratio	0	0.12
Substrate temperature	640 °C	640 °C
Growth time	5 min	55 min

growth resumed between each section for 60 s. This series of 5 QDs was repeated 5 times, and more about the structure is shown in Chap. 5.

2.2.2 *Microtome*

A Leica EM UC7 ultramicrotome is used to produce cross-sections of NWs. The process is illustrated in Fig. 2.5 with the steps as follows. A section of substrate with NWs attached is taken (a) and the NWs are embedded using a low-viscosity resin (b). This is submerged into liquid nitrogen which causes the substrate to break away from the embedded NWs (c). The embedded NWs are rotated, with the bottom of the NWs exposed at the top of the resin (d), and this is shaped using a shaping microtome knife leaving a trapezoid shape (e). With the cutting knife and sample aligned, slices are cut with the embedded NWs moved in a repeated downwards motion (f). The slices are cut into a small bath of de-ionised water where the microtomed slices float ((g) and (h)) and make for easy transfer to a TEM grid.

An example photo of some microtomed slices floating in the water bath are shown in Fig. 2.6a. These slices are transferred to a TEM grid by gently applying a TEM grid to the floating slices, with the same transferred slices shown in Fig. 2.6b. A series of stitched SE SEM images of the TEM grid with microtome slices is shown in Fig. 2.6c.

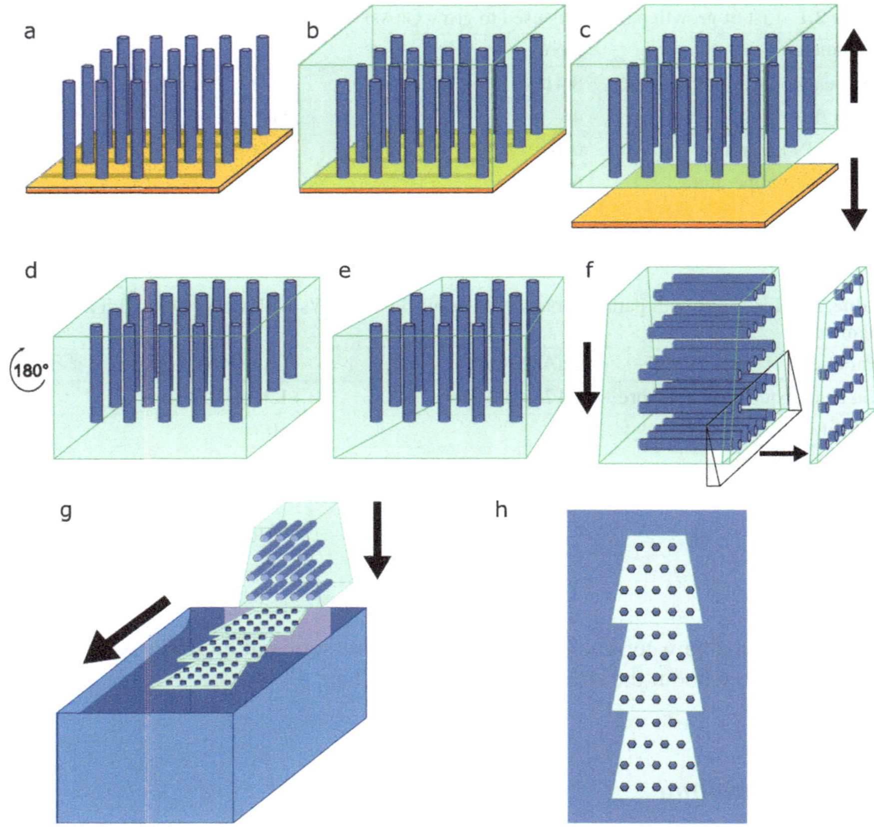

Fig. 2.5 Using a microtome to slice NWs. **a** NWs attached to substrate. **b** NWs embedded in low-viscosity resin. **c** Substrate removed using liquid nitrogen. **d** Resin block rotated to show exposed NW tips at the top. **e** Resin is shaped into a trapezoid using a diamond shaping knife. **f** Sample is sliced using a static diamond knife by pushing the sample down the knife. **g** As the resin block is sliced the slices collect in a water bath and float ready for collection and transfer to a TEM grid. **h** Typical appearance of a ribbon of multiple sequential slices

2.2.3 Conventional TEM and STEM Preparation

NWs for conventional imaging in a TEM are simple to prepare. Transferring NWs from substrate to TEM grid simply requires mechanical scraping of the copper holey carbon grid across the substrate with NWs attached using tweezers. This is illustrated in Fig. 2.7.

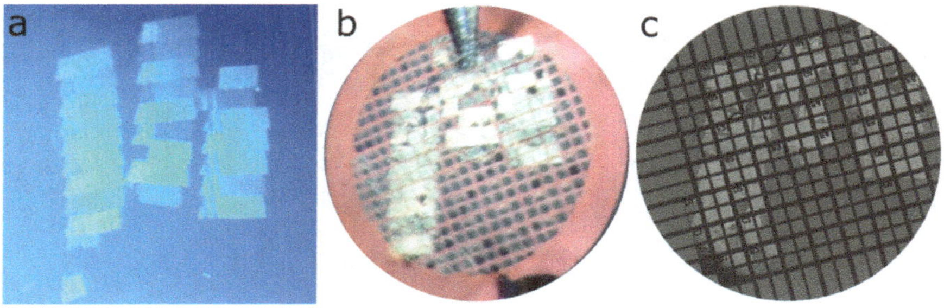

Fig. 2.6 **a** Fresh microtome slices in the water bath ready to be transferred to a TEM grid. **b** The same microtome slices transferred to a TEM grid. **c** A series of SE SEM images stitched together of the TEM grid with microtome slices

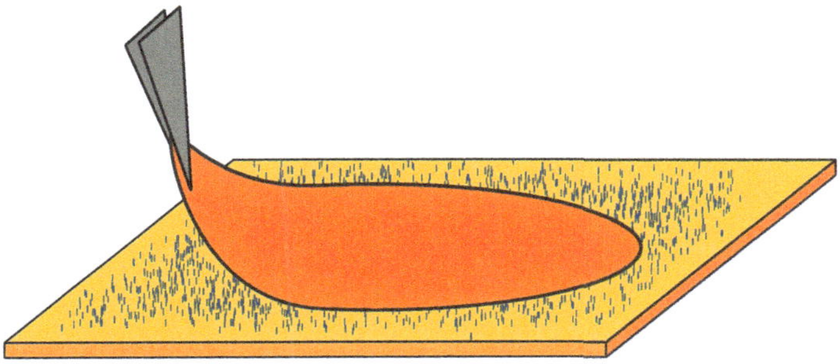

Fig. 2.7 Mechanical scraping with tweezers to transfer NWs from substrate to a copper TEM grid

2.2.4 *In-Situ Microscopy*

In-situ microscopy was performed using a DENSsolutions Wildfire chip [9] in a double tilt holder. These are Si_3N_4 chips with electron transparent windows for samples to be placed and a heating element that allows for temperatures of up to 1100°C with very small sample drift. Figure 2.8a, b show SE SEM images of a Wildfire chip before adding the sample. The light spiralled contrast is the heating element and the dark contrast regions are the electron transparent windows.

Two methods were used to add NWs to the chip, the first of these being the ethanol dropcast technique. For this method, a small piece of the substrate with NWs attached was cleaved and placed in ethanol. These were sonicated using a sonic probe for 10 min and then pipetted onto the chip. The chips were checked for NW distribution using an optical microscope, and if more NWs are needed the process is repeated. The second method uses a lint-free tissue, and involves first gently wiping a small section of tissue onto the substrate with NWs attached to transfer them to the tissue, and then applying this tissue to the chip using very light force. This technique reduces the amount of contamination compared to the ethanol dropcast technique.

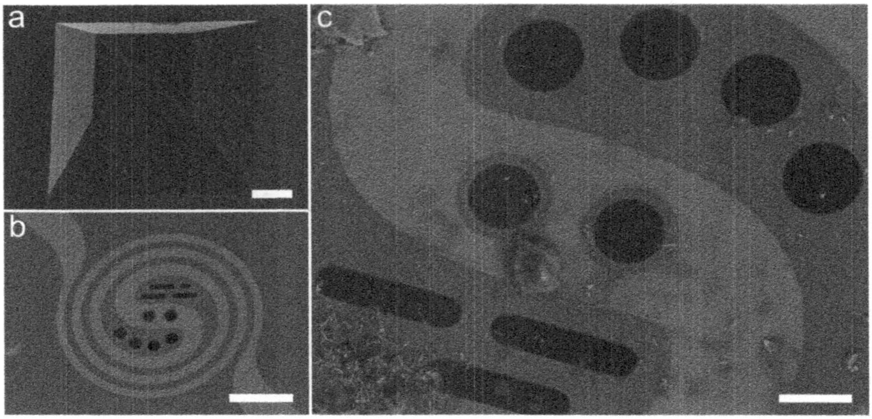

Fig. 2.8 A series of SE SEM images of a DENS solutions Wildfire Si_3N_4 chip at increasing magnification. The spiral pattern is the heating element and the dark circles are electron-transparent windows. **a**, **b** show a clean chip and **c** has NWs dispersed on the chip. Accelerating voltage 10 kV. Scale bars are **a** 200 μm **b** 50 μm **c** 10 μm

2.3 Simulations

High resolution simulations of TEM and STEM adopt one of two approaches, either Bloch wave propagation or multislice principle. Multislice is the most commonly used approach [1] since Bloch wave propagation is computationally more expensive [2]. The multislice technique was first derived by Cowley and Moodie in 1957 [10]. In multislice simulations, the sample is first sliced into thin sections perpendicular to electron beam direction. Each slice is projected onto a plane with a projected potential for the slice. The amplitude and phase of electron beams generated by the interaction of the incident beam with the projection plane is calculated. These beams then propagate through free space inside the microscope until they meet the next slice and its projected potential. The scattering calculation is repeated for all incident electron beams, which produces a new set of beams that go on to propagate to the next slice [11]. For TEM the initial wave is a plane wave and for STEM the initial wave is a probe. For TEM images, the end process of the simulation is to take the wave function after it has propagated all the way through the sample, then apply the transfer function of the objective lens to get the image wave function and then take the square modulus of the image wave function to get image intensity. For STEM images, the probe wave function is propagated through the sample slices until it exits the sample, at which point the transmitted wave function is Fourier transformed to get the wave function in the diffraction plane. The square modulus of the wave function in the diffraction plane is integrated over the range of the detector being used which then gives the signal for one pixel in the final image. This process is repeated for each pixel in the image.

Table 2.3 Mean-square displacements calculated using values reported by Schowalter et al. [12]

Material	Atom	A	B	σ	$\langle u^2 \rangle (\text{Å}^2)$
GaAs	Ga	4.155864e+12	2.035401e+13	44.79912	8.6308e-3
	As	4.774990e+12	2.119684e+13	45.44807	7.4201e-3
GaP	Ga	3.118889e+12	2.273464e+13	51.18278	6.9568e-3
	P	1.035200e+13	3.687135e+13	72.62717	6.2241e-3

An important factor that must be considered particularly for STEM imaging is TDS. This is included in image simulation by offsetting the position of atoms by a small random amount, performing simulations for a series of positions and finally an image is built from the average of several configurations. This technique is known as the frozen phonon approach [11]. Simulations performed in this work use the mean-square displacements $\langle u^2 \rangle$ (Å^2) for temperature $T = 293$ K calculated using the equation reported by Schowalter et al. [12]

$$\langle u^2 (\nu; T) \rangle = \frac{\hbar}{2M_\nu} \frac{\coth\left\{ \frac{\hbar\left[A \exp\left(\frac{-T^2}{\sigma^2}\right)+B\right]}{2k_B T} \right\}}{\left[A \exp\left(\frac{-T^2}{\sigma^2}\right) + B\right]} \tag{2.18}$$

where ν is the element, M_ν is the mass of atom ν, and A, B, and σ are fit parameters. This is used when TDS is included in simulations, and calculated values of $\langle u^2 \rangle$ for elements relevant to this work are listed in the final column of Table 2.3 with the other columns listing parameters used to obtain these values, as reported by Schowalter et al. [12]. Image simulations in this thesis were produced using clTEM software [13, 14]. Unless otherwise stated, simulated images are produced using voltage of 200 kV, aperture 24 mrad, and spherical aberration coefficient 1 μm.

2.4 Measuring Strain Using GPA

A commonly used technique to determine strain of a sample from a TEM/STEM image is GPA [15, 16]. The GPA technique combines real space and reciprocal space information. The technique uses the idea that a real image of a perfect crystal $I(r)$ can be described by a fourier series [15]

$$I(r) = \Sigma g H_g \exp\{2\pi i g \cdot r\} \tag{2.19}$$

where r is position, g is the Bragg reflection, and H_g are the Fourier coefficients given by

$$H_g = A_g \exp\{i P_g\} \tag{2.20}$$

where A_g is the amplitude of lattice fringe, and P_g is the phase, which gives positions of fringes in the image.

Using the GPA technique first requires two non-colinear reciprocal lattice vectors to be chosen from the FFT of the image and masked in fourier space. The inverse fourier transform of this gives a complex image and from this image the bragg-filtered image intensity, the amplitude, and the phase of the image can be calculated. The calculated phase image gives the component of the displacement field in the direction of the reciprocal lattice vector g chosen. By combining the information from two sets of lattice fringes, a vectorial displacement field can be calculated, and finally the gradient of the displacement field gives lattice distortion, and from this strain can be calculated.

Care must be taken when using this technique that appropriate lattice vectors are chosen, as for example in twinned materials, some g vectors can incorrectly show strain where there is none. Artefacts can be avoided by choosing g vectors such that $g \cdot v = n$ where v is the lattice displacement vector between the two lattices involved in the structure and n is an integer [16].

2.5 Spectroscopy

2.5.1 Cathodoluminescence

CL in a SEM is a local luminescence technique which involves the analysis of light emission from solids under the excitation of an electron beam. An incident electron beam, with energies that can range from a few hundred eVs to tens of keVs, impacts on the sample surface and a product of this interaction is the generation of electron-hole (e-h) pairs through the impact ionisation mechanism [5, 17, 18]. The energy of the incident electrons is normally much greater than the bandgap of the material and results in a carrier generation rate greater than the incident electron flux. The e-h pairs generated can recombine which can either be radiative or non-radiative. The radiative recombination of e-h pairs produces photons with system specific wavelengths and this is the signal that is measured in CL.

There are a few different possible mechanisms that can contribute to the emitted light, each producing different wavelengths of light. The first of these is spontaneous emission from band-to-band transitions, with the energy of these photons given by the material bandgap E_g. Free excitons can also contribute to emission. With free excitons treated as hydrogen atoms, then they have a series of excited states, with emission lines given by [18]

$$E(n) = E_g - \frac{1}{n^2} E_x \tag{2.21}$$

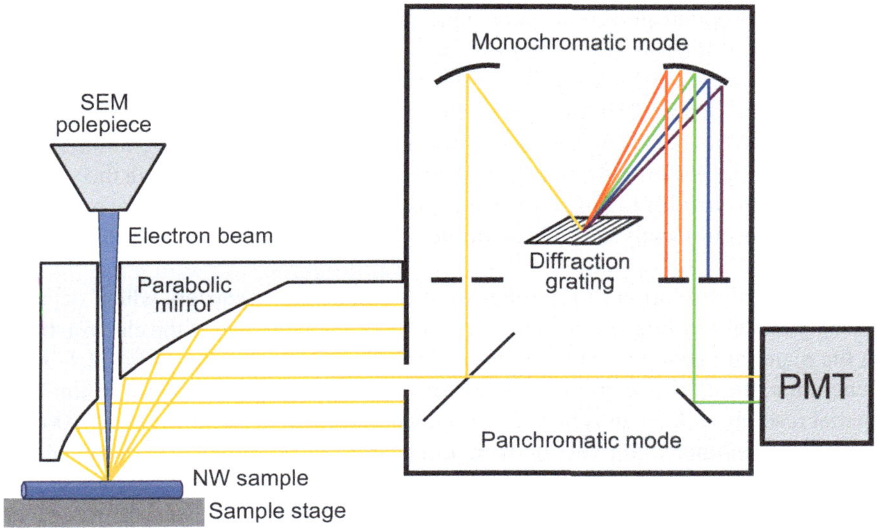

Fig. 2.9 Measuring a CL signal from a sample inside a SEM. A parabolic mirror allows the electron beam through a small aperture in the mirror. The mirror directs light emitted by the sample towards a PMT. The light is either sent directly to the PMT which is known as panchromatic mode, or the light is directed to a series of mirrors and a diffraction grating that splits the light into its constituent wavelengths. This allows for the selection of which wavelength of light is sent to the PMT and measured. This mode of operation is known as monochromatic mode

where n is the principle quantum number and E_x is the binding energy. In the case of structures with quantum confinement, the binding energy is increased. Another possible influence on emission is from any defects that may be present. Defects, like point defects, can introduce additional defect states whose energy lie within the bandgap and can lead to emissions with lower energy. The spectral peaks from the different contributions are broadened with temperature, with a full width at half max $\approx 2k_bT$ [18]. The use of a cryogenic stage can help to narrow and resolve some emission peaks.

In terms of how the equipment operates, there are three main stages. First is the collection of the light emitted from the sample, which from SCs is usually dispersed because of a large refractive index [18]. Light collection is usually done with parabolic or elliptic mirror, which have an aperture through which the electron beam is directed. An example of a CL system is shown in Fig. 2.9, and demonstrates this. The collected light is then guided to a light detection system, and usually takes one of two paths. One path is to direct all the collected light directly to the detector (the yellow line in Fig. 2.9). In this mode, all wavelengths are included in the measured signal and is known as panchromatic mode. Alternatively the collected light can be guided to a diffraction grating which then splits the light into its constituent wavelengths, and if a small spectral window is used a limited range of wavelengths can be chosen and measured. This mode is known as monochromatic mode. The selected light is then sent to an appropriate photon detector, which should be chosen in accordance to what wavelengths of interest are being measured. For UV light and visible light, PMTs are typically used which are sensitive to the range 300–1700 nm [18]. For light in the infrared region of up to 2100 or 1800 nm then InGaAs or Ge detectors respectively can be used [18]. With both panchromatic and monochromatic modes of operation, spatial mapping can be performed to analyse where emission is coming from in the sample, with monochromatic allowing for wavelength selectivity. Monochromatic mode also allows for spectral analysis, and can probe how intensely different wavelengths are being emitted.

The spatial resolution of CL relies on the generation of carriers which depends strongly on the accelerating voltage, with the interaction volume of the electron beam in the material varying strongly with accelerating voltage, as seen in Fig. 2.4. After generation, the carriers can diffuse before recombination, and both of these can limit spatial resolution. CL can achieve a spatial resolution down to below a few 100 nm. The change in interaction volume with accelerating voltage means that some depth sensitivity is possible with CL.

CL mapping in Chap. 3 was acquired using a Gatan MonoCL3 spectrometer with sample cooled to $\approx -100\,^{\circ}$C. CL spectra were acquired using a spectral resolution of 2 nm.

2.5.2 Energy Dispersive X-Ray Spectroscopy

Inside a TEM or SEM the incident electrons may interact with the sample inelastically and lose energy to the sample. This type of scattering produces a range of signals, and one of these is characteristic x-rays [1]. When a high-energy electron interacts with an atom, it can penetrate the outer bound electron cloud, and then interact with the more tightly bound core shell electrons. If enough energy is transferred, then a core electron is ejected and leaves a hole. This leaves the atom in an excited state, which can return to a lower energy state by filing the hole with an electron from the outer shell. This transition is accompanied by the emission of an x-ray or Auger electron, and the energy of these is characteristic of the energy difference between the two electron shells involved in the transition and is unique to the atom. If the incident electrons penetrate through the core electrons, they can interact inelastically with the nucleus. By interacting with the Coulomb field of the nucleus, a change in momentum can occur and during this process it can emit x-rays, known as bremsstrahlung radiation. Since the electron can lose any amount of energy via this mechanism, these x-rays can have any energy up to the beam energy and have a continuous energy spectrum which superimposes on top of the characteristic x-rays. To collect this signal, a detector inside the TEM is required, and the most commonly used is a Si drift detector [1]. These detectors work by generating a charge pulse proportional to the x-ray energy, with very rapid processing of individual x-rays done to build a spectrum. A spectrum is built by separating the x-rays into their energies and counting how many are detected over an energy range. Example spectra can be found in [1] and more information on TEM sample interaction can be found in a review by Lorimer [19]. Spectra acquisition can either be done for a single point on the sample, i.e. with the electron beam condensed and probing one spot on the sample, or maps can be built by acquiring EDX spectra for a series of points on the sample. This can be used to provide elemental mapping, with atomically resolved elemental mapping possible [20, 21].

In order to quantify the composition based on EDX intensities, the Cliff-Lorimer ratio technique is used [22]. In a binary system, the concentration of species A and B (C_A, C_B) and the corresponding intensities (I_A, I_B) are related by the equation [22]

$$\frac{C_A}{C_B} = k_{AB}\frac{I_A}{I_B} \tag{2.22}$$

where k_{AB} is the Cliff-Lorimer factor, or the k-factor. The k-factor is different for each pair of elements and depends on other factors such as detector efficiency. To obtain a value for C_A and C_B it is assumed that elements A and B make up the specimen and so

$$C_A + C_B = 100\% \tag{2.23}$$

These equations can be extended to higher order systems, e.g. for a ternary system the following equation can be used

$$\frac{C_B}{C_C} = k_{BC}\frac{I_B}{I_C} \tag{2.24}$$

$$C_A + C_B + C_C = 100\% \tag{2.25}$$

k-factors can also be calculated from first principles and are given by the equation [1]

$$k_{AB} = \frac{(Q\omega a)_B\, A_A}{(Q\omega a)_A\, A_B} \tag{2.26}$$

where Q is ionisation cross section, ω is fluorescence yield, a relative transition probability and A is atomic weight. In this thesis, EDX is quantified using calculated Cliff-Lorimer k-factors from Oxford Instruments Aztec software.

References

1. Williams D, Carter C (2009) Transmission electron microscopy a textbook for materials science. Springer
2. Brydson R (2011) Aberration-corrected analytical transmission electron microscopy. Wiley
3. Pennycook S, Nellist P (2011) Scanning transmission electron microscopy imaging and analysis. Springer
4. Jones L et al (2015) Smart Align-a new tool for robust non-rigid registration of scanning microscope data. Adv Struct Chem Imaging 1:8

5. Ul-Hamid A (2018) A beginners' guide to scanning electron microscopy. Springer
6. Beyer A et al (2017) Influence of surface relaxation of strained layers on atomic resolution ADF imaging. Ultramicroscopy 181:8–16
7. Zhang Y et al (2013) Self-catalyzed GaAsP nanowires grown on silicon substrates by solid-source molecular beam epitaxy. Nano Lett 13:3897–3902
8. Zhang Y et al (2016) Influence of droplet size on the growth of self-catalyzed ternary GaAsP nanowires. Nano Lett 16:1237–1243
9. DENS Solutions Wildfire Nano-chip https://denssolutions.com/products/wildfire/nano-chip/. (Accessed: 16/05/2020)
10. Cowley JM, Moodie AF (1957) The scattering of electrons by atoms and crystals a new theoretical approach. Acta Crystallogr 10:609–619
11. Kirkland E (2010) Advanced computing in electron microscopy. Springer
12. Schowalter M, Rosenauer A, Titantah JT, Lamoen D (2009) Computation and parametrization of the temperature dependence of Debye-Waller factors for group IV, III-V and II-VI semiconductors. Acta Crystallogr Sect A 65:5–17
13. clTEM https://jjppeters.github.io/clTEM/. (Accessed: 16/05/2020)
14. Peters JJ (2021) JJPPeters/clTEM: v0.3.4 version v0.3.4
15. Hÿtch M, Snoeck E, Kilaas R (1998) Quantitative measurement of displacement and strain fields from HREM micrographs. Ultramicroscopy 74:131–146
16. Peters JJ et al (2015) Artefacts in geometric phase analysis of compound materials. Ultramicroscopy 157:91–97
17. Edwards PR, Martin RW (2011) Cathodoluminescence nano-characterization of semiconductors. Semicond Sci Technol 26:064005
18. Jimenez J, Tomm J (2016) Spectroscopic analysis of optoelectronic semiconductors. Springer
19. Lorimer GW (1987) Quantitative X-ray microanalysis of thin specimens in the transmission electron microscope; a review. Mineral Mag 51:49–60
20. Kothleitner G et al (2014) Quantitative elemental mapping at atomic resolution using x-ray spectroscopy. Phys Rev Lett 112:085501
21. Mostaed A, Balakrishnan G, Lees MR, Beanland R (2018) Electron-irradiation induced defects in Yb2Ti2.05O7. Acta Materialia 143:291–297
22. Cliff G, Lorimer GW (1975) The quantitative analysis of thin specimens. J Microsc 103:203–207

Chapter 3
Defects in Nanowires

3.1 Introduction

As seen in Sect. 1.3.1, one of the attractive properties of NWs is the physical size and morphology of NWs results in fewer defects. This characteristic comes from more efficient relaxation of strain and image forces from NW surfaces tending to remove dislocations, minimising NW energy. While NWs have a reputation for being less defective than their thin film and bulk counterparts, defects can and do exist.

The most common defect in III-V NWs is the ortho-twin, and these have been widely reported in the literature [1–5]. The formation of ortho-twins is relatively well understood, as was seen in Sect. 1.3.1. These twins are perpendicular to the NW growth direction, which is usually $[\bar{1}\bar{1}\bar{1}]$. A less common twin in NWs is the $\Sigma3\{112\}$ twin, and provided polarity is conserved along the NW axis then this is a para-twin. This type of twin has been seen previously in other material systems including Cu [6, 7], Au [8], β-SiC [9], diamond [10], Si [11–14], III-Vs [15], and II-VIs [16, 17]. $\Sigma3\{112\}$ twins were recently found in self-catalysed GaAsP NWs with sub-optimal growth conditions [18]. These defects cover the majority of previously reported defects in NWs.

This chapter shows that a surprising variety of defects can be found in NWs and atomic resolution STEM images are presented for each type. The defects are classified by their Burgers vector, and an extensive number have been studied to find out how common each variety is in the sample examined. This chapter also provides evidence of how the defects form during growth by using in-situ heating experiments.

3.2 Types of Defects in GaAs(P) Nanowires

The GaAs(P) NW system has proven to be an unexpectedly rich system in terms of the type of defects that can be found. Figure 3.1 shows a selection of low magnification BF-STEM images of defective NW tips where the many lines of contrast suggest

J. A. Gott, *Defects in Self-Catalysed III-V Nanowires*, Springer Theses,
https://doi.org/10.1007/978-3-030-94062-1_3

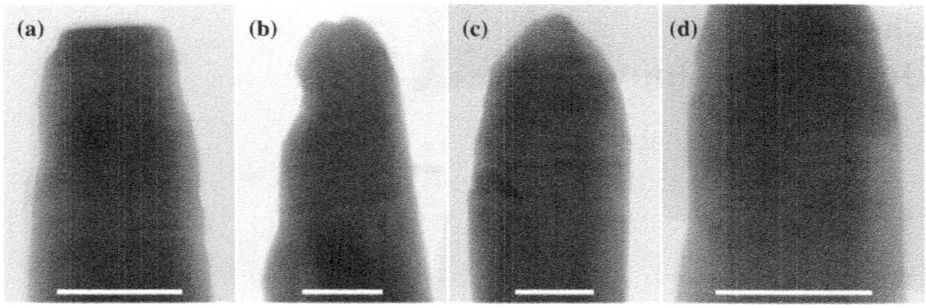

Fig. 3.1 **a–d** BF-STEM images of defective NW tips. Dark lines of contrast indicate the presence of defects Scale. The shape of NW tips in (**b**) and (**c**) are more tapered than those in (**a**) and (**d**). Scale bars 100 nm

these NW tips are not perfect ZB crystals. BF-STEM images like these provide better contrast (compared to HAADF-STEM) to show defect features at low magnification. Here, atomic resolution STEM images of a number of self-catalysed GaAsP NWs shows that a variety of line defects with zero Burgers vector can exist in NWs. Additionally the GaAsP NW system has shown a wealth of defects with non-zero Burgers vector. These defects have been found where multiple twinning occurs and produces core configurations that are not commonly seen.

Evaluating the Burgers vector of a defect is of interest as it indicates if it has a long range strain field. Burgers vector analysis is not straight forward when multiple twinning is present. The approach used here takes a chosen sublattice (either group III or V) and draws a closed circuit around the defect using lattice vectors, while minimising the component out of the image plane. The process is outlined in more detail in the following section.

3.2.1 Analysing Burgers Vectors

The conventional method of finding Burgers vector for a dislocation was outlined in Sect. 1.3. This process becomes difficult as soon as any type of twinning is involved and even more so for defects in multiply twinned regions. In the $\langle 110 \rangle$ projection of ZB materials, the group III and V sublattices are displaced from each other by a vector $\frac{1}{4}[001]$, which gives the appearance of a pair of atoms known as a dumbbell. In this approach to finding the Burgers vector for defects in twinned ZB crystals, one sublattice must be chosen and used consistently. Either sublattice can be chosen when constructing the circuit, as long as the same is used throughout. A clockwise closed circuit is drawn around the defect using lattice vectors. The Burgers vector is the negative sum of the lattice vectors that make up the closed circuit. The most likely Burgers vector will be given by the sum of lattice vectors that has the smallest magnitude, and so an appropriate choice should be made between vectors with

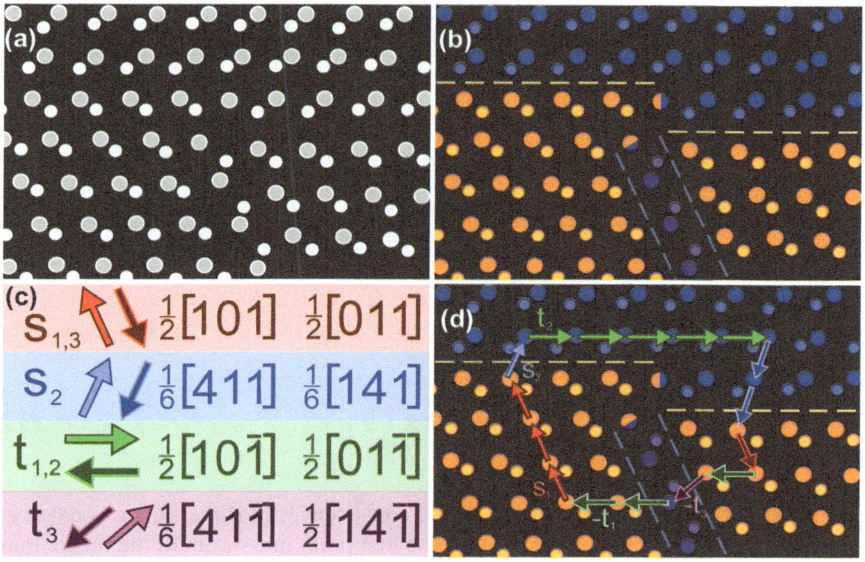

Fig. 3.2 **a** An illustration of typical dumbbell appearance in ADF-STEM images of a twinned ZB structure observed in a ⟨110⟩ direction. **b** The same image with dumbbells coloured according to crystal orientation. The orange dashed lines indicate (111) twin boundaries and the blue dashed lines indicate $(11\bar{1})$ twin boundaries. Line direction into the image $\left[\bar{1}10\right]$. **c** Lattice vectors to describe translations in different frames of reference. The light fill/dark border are *positive* vectors while the reverse are *negative* vectors. The vectors are grouped (*left/right*) according to having a vector component into (*left*) or out (*right*) of the image. **d** Using the vectors in (**c**) to describe a circuit around the defect structure in (**b**)

components into or out of the image when summing vectors, and the component perpendicular to the image should be minimised. Where twins are involved, a single crystal reference frame must be chosen and vectors from twinned crystals must be described using the chosen reference frame.

To demonstrate this approach, an example of a multiply twinned structure is shown in Fig. 3.2. Figure 3.2a represents the structure as would be seen in an ADF-STEM image. To make clear the different orientations of the material in the area being analysed, the same structure is shown in Fig. 3.2b with dumbbells coloured differently for each orientation. The twin planes are indicated by dashed lines. With line direction into the image $[\bar{1}10]$, Fig. 3.2c lists the relevant vectors for this example, with vectors on the left having components into the image and those on the right with components out of the image. To obtain the Burgers vector **b**, as mentioned previously, all vectors used must be written in the same reference frame. Some vectors are no longer lattice translation vectors when expressed in a different reference frame. When a dislocation lies in an interface, there will be two equivalent ways to describe the same dislocation, depending on which reference frame is chosen. In Fig. 3.2 there are three different reference frames: orange atoms (frame 1), blue atoms (frame 2) and purple atoms (frame 3). In the vectors used in the analysis, s_i and t_i, the subscript refers to the reference frame. In this example s_1 and t_1 are the basis vectors describing the lattice in frame 1. To write the basis vectors of frame 2 into reference frame 1, a transformation P is applied, in this case a 180° rotation about [111],

$$P = \frac{1}{3}\begin{bmatrix} \bar{1} & 2 & 2 \\ 2 & \bar{1} & 2 \\ 2 & 2 & \bar{1} \end{bmatrix} \tag{3.1}$$

To write the vectors of frame 3 in reference frame 1, a 180° rotation about $[11\bar{1}]$ (which is the direction for the blue dashed twin plane in Fig. 3.2b) is applied using

$$Q = \frac{1}{3}\begin{bmatrix} \bar{1} & 2 & \bar{2} \\ 2 & \bar{1} & \bar{2} \\ \bar{2} & \bar{2} & \bar{1} \end{bmatrix} \tag{3.2}$$

The vectors t_1 and t_2 are equivalent. To be consistent with three-dimensional crystallography, lattice vectors with components that point into or out of the image must be used, even if these components parallel to the electron beam are not visible. For each lattice vector here, there are two possible vectors because the component of vector that points into/out of the image cannot be detected when viewing a projected image. When using these vectors to construct a circuit, one lattice direction is chosen to be positive. In the example shown here, the positive and negative vector directions are shown by a difference in arrow colour for each frame. The *positive* direction has arrows with light colour fill/dark outline while *negative* directions have a dark colour fill/light outline.

With the dislocation line direction into the image, the Burgers vector **b** is given by the negative sum of the vectors that make up the circuit. The circuit for the example shown in Fig. 3.2d starting at the bottom left corner is then given by

$$-\mathbf{b} = 3s_1 + s_2 + 5t_2 - 2s_2 - s_1 - t_1 - t_3 - 2t_1 = 2s_1 - s_2 + 2t_1 - t_3 \tag{3.3}$$
$$-\mathbf{b} = \frac{2}{2}[101] - \frac{1}{6}[411] + \frac{1}{2}[10\bar{1}] + \frac{1}{2}[01\bar{1}] - \frac{1}{6}[41\bar{1}] = \frac{1}{6}[110]$$

One thing to note here is the choice of each vector. Since there are two possible choices for each vector, and we cannot determine the component into or out of the image, the choice is made such that the net out-of plane component of the circuit in the image is minimised.

In the following sections, this procedure is used to identify the Burgers vector of a wide variety of defects found in self-catalysed III-V NWs. It soon becomes apparent how this method makes Burgers vector analysis much easier where the material is multiply twinned.

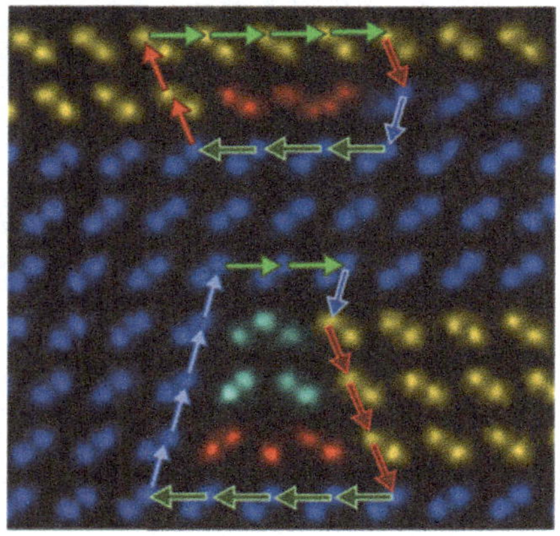

Fig. 3.3 Coloured band-pass filtered ADF-STEM image of a 1ML intrinsic partial dislocation (upper) and a 3ML defect (lower). The red colouring indicates a 1ML intrinsic partial dislocation structure and the cyan colouring indicates a 2ML extrinsic partial dislocation structure. Each defect has a vector circuit drawn around it to determine **b**. The image is 3 nm wide

3.2.2 Σ3{112} *Defect—The Three Monolayer Defect*

The first type of defect to be examined is the Σ3{112} twin boundary, with an example of one shown at the bottom of Fig. 3.3. This Σ3{112} twin boundary defect was previously reported by Sanchez et al. in self-catalysed GaAsP NWs [18] and they have also been reported by Zamani et al. in self-catalysed GaAs NWs [19]. This defect will be referred to as the 3 monolayer (ML) defect in reference to the number of MLs that form the twin interface, with a ML consisting of a dumbbell pair.

The 3ML defect can be thought of as being built from a combination of a 1ML and a 2ML $\frac{1}{6}\langle 112\rangle$ intrinsic and extrinsic partial dislocation with opposite **b**, or in other words the 3ML defect is essentially a dipole of two partial dislocations. The example in Fig. 3.3 indicates the two partials by adding colour to the defect, with the 1ML component in red and the 2ML component in cyan. Circuits used to calculate **b** are shown in Fig. 3.3 and the vector sum for the 3ML defect is given by

$$-\mathbf{b} = 2t_2 - s_2 - 3s_1 - 4t_2 + 4s_2 = 3s_2 - 3s_1 - 2t_2 \tag{3.4}$$
$$-\mathbf{b} = \frac{3}{6}[411] - \frac{3}{2}[101] - \frac{1}{2}[10\bar{1}] - \frac{1}{2}[01\bar{1}] = 0$$

and so the 3ML defect has a Burgers vector of zero. For the 1ML defect at the top of Fig. 3.3 the vector sum is given by

$$-\mathbf{b} = 4t_1 - s_1 - s_2 - 3t_1 + 2s_1 = t_1 + s_1 - s_2 \tag{3.5}$$
$$-\mathbf{b} = \frac{1}{2}[10\bar{1}] - \frac{1}{2}[101] - \frac{1}{6}[411] = \frac{1}{6}[2\bar{1}\bar{1}]$$

which is the Burgers vector for a 30° Shockley partial dislocation.

In order to check for strain, GPA analysis is performed on the image (see methods Sect. 2.4) and is shown in Fig. 3.4. Figure 3.4a shows the image used, and Fig. 3.4b

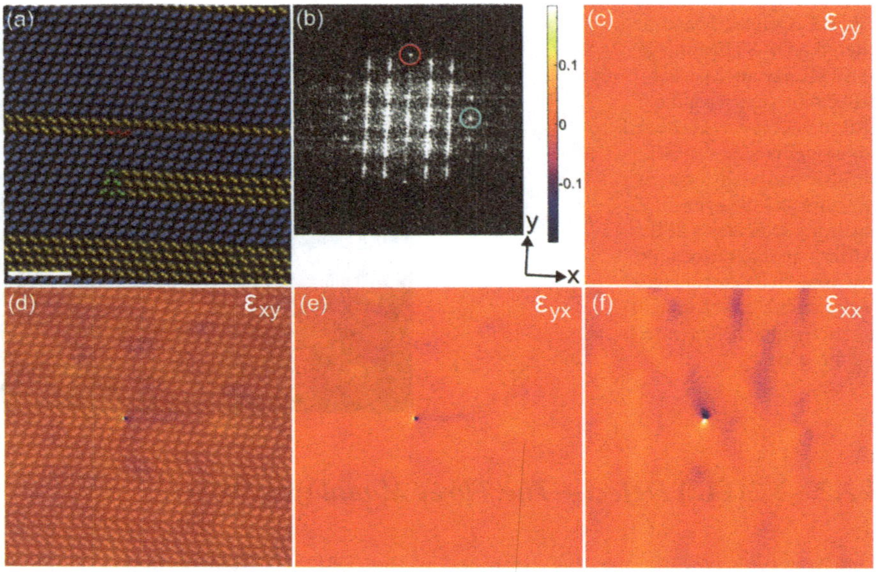

Fig. 3.4 **a** Coloured band-pass filtered ADF-STEM image with a 1ML intrinsic partial (red) and a 3ML defect (green). Scale bar 2 nm. **b** The FFT of the image in (**a**) with the reflections chosen for GPA analysis marked by red and blue circles. The scale bar for strain maps is shown here with the *x* and *y* directions used in the strain calculation indicated below. **c–f** Strain maps for the corresponding components labelled in each image, where **d** has an overlay of the strain map and ADF image

shows the FFT of Fig. 3.4a and a red and a blue circle indicate which spots are used to perform GPA analysis. Twinned regions are present in the image and so spots are chosen that are common to both crystal orientations, with this shown more clearly in Fig. 1.7 in introduction Sect. 1.3. The *x* and *y* directions are indicated by the axis at the bottom right of Fig. 3.4b and the strain colour scale is shown on the right side of Fig. 3.4b. The strain component maps in Fig. 3.4c–f indicate no strain associated with the 3ML defect while there is strain associated with the 1ML Shockley partial dislocation, which is most easily seen in Fig. 3.4d where the ϵ_{xy} strain map is overlaid on top of the image. This GPA analysis therefore confirms there is no long-range strain associated with the $\mathbf{b} = 0$ 3ML defect.

Another property of significance is the variety of different configurations that are possible for the 3ML defect. As mentioned in the introduction the 3ML defect can either act as a step in a twin boundary, or they can appear as a section of twinned material. It has been observed in multiple NWs that steps or twinned sections of material can be 6ML or more high. There are also examples where a 3ML type defect is adjacent to a partial dislocation. The 3ML defect causes no additional strain on the partial dislocation as was seen in Fig. 3.4. After an extensive study into the types of defect found in NWs in which over 300 defects were analysed, the 3ML type defect is by far the most common type of defect observed. The image in Fig. 3.5

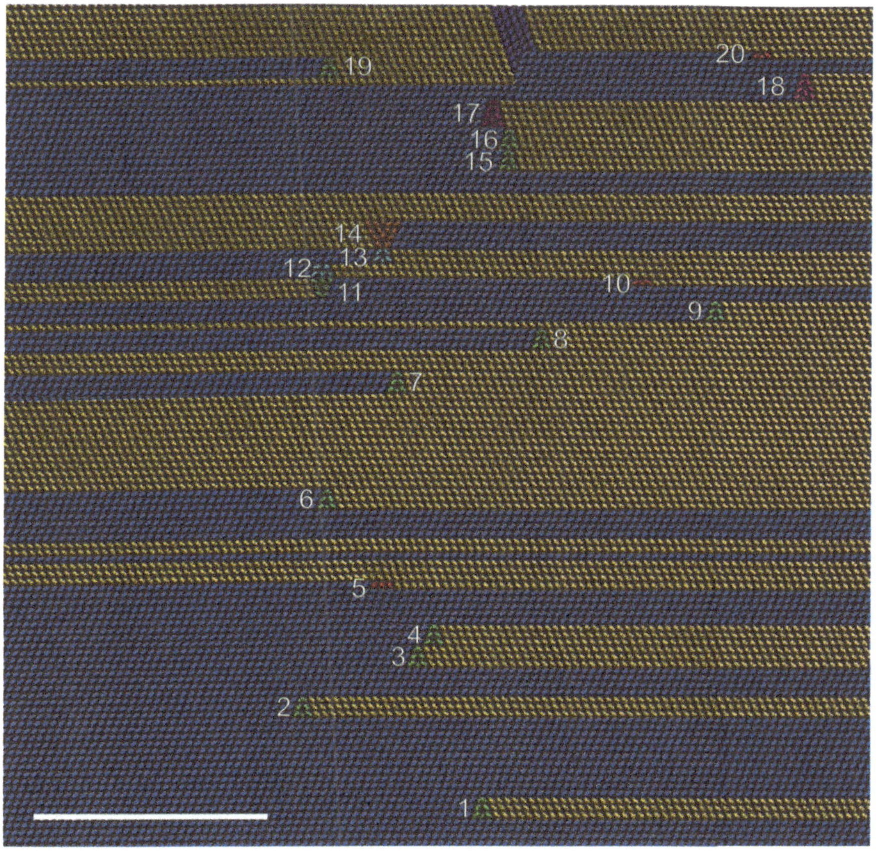

Fig. 3.5 Coloured band-pass filtered ADF-STEM image of an area of a NW tip with defect structures labelled with numbers. Of the observed defects, many of them are 3ML type defects (green). Other defects observed include 1ML defects (red) and 2ML defects (cyan). Unusual examples of 4ML defects are seen at defects labelled 14, 17 and 18. Scale bar 10 nm

is an atomic resolution ADF-STEM image of an area of a NW tip containing many defects. The defect structures have been coloured according to the number of affected MLs, with 1ML in red, 2ML in cyan, and 3ML in green. Two unusual 4ML structures are coloured in pink and orange.

20 defects are labelled in Fig. 3.5 with some defects having dislocation character, and 12 of the labelled defects are of the 3ML type which demonstrates how common the 3ML type defect can be. Figure 3.5 shows examples of the different configurations of the 3ML defect, and include:

- Twinned section of material extending to the right (defects 1 and 2)
- Twinned section of material extending to the left of the image (defects 7, 8 and 19)

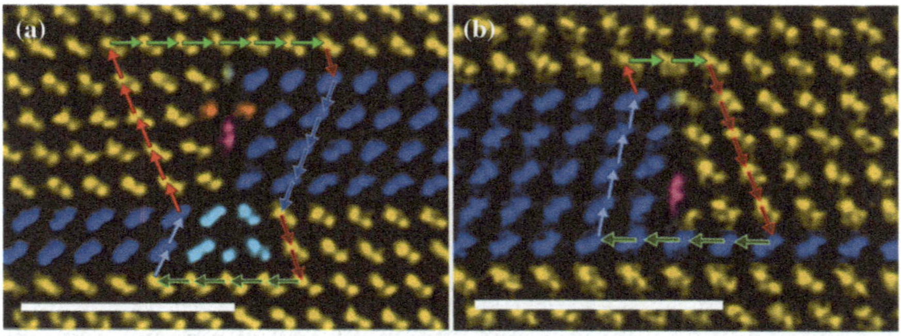

Fig. 3.6 Coloured band-pass filtered ADF-STEM images of specific defects from Fig. 3.5. **a** shows defects 13 and 14, and **b** shows defect 18. Circuits are drawn around each for Burgers vector analysis. Unusual core structures are highlighted by the orange and pink atoms. Scale bars 2 nm

- Twinned section with multiple 3ML defects (defects 3 and 4 and 15 and 16)
- A (111)-twin down-step (defect 6)
- A (111)-twin up-step (defect 9)

Other defects observed include an example of a twin extending across the entire image width (between defects 5 and 6). Examples of intrinsic 30° partials can be seen with defects 5, 10 and 20, and examples of 30° extrinsic partials can be seen with defects 12 and 13. Defects 14, 17 and 18 are 4 MLs in height and their core structure is unusual. Typical defects observed that are 4MLs high would usually consist of a 3ML and a 1ML defect (an example of this can be seen by defects 1 and 2 in Fig. 3.13). A closer view is shown in Fig. 3.6. Figure 3.6a has an example with two isolated single atom columns (highlighted in orange) and a pair of atom columns aligned vertically (highlighted in pink). Isolating this 4ML defect is difficult here since there appears to be a 2ML extrinsic partial below this. A circuit is drawn around the complete 6ML structure and can be described by the sum

$$-\mathbf{b} = 6t_1 - s_1 - 4s_2 - 2s_1 - 4t_1 + 2s_2 + 5s_1 = 2s_1 - 2s_2 + 2t_1 \qquad (3.6)$$
$$-\mathbf{b} = \frac{2}{2}[101] - \frac{2}{6}[411] + \frac{1}{2}[10\bar{1}] + \frac{1}{2}[01\bar{1}] = \frac{1}{6}[11\bar{2}]$$

Which is the Burgers vector of a Shockley partial dislocation. This structure could be described as a group of three 2ML defects, where two lower 2ML defects form a dipole whose Burgers vectors cancel each other, with the upper 2ML contributing to the net Burgers vector measured. This arrangement is not easy to identify because of the unusual core structure. Figure 3.6b has an example that looks to be a combination of an upper 3ML and a lower 1ML defect interacting, with the resulting structure having an unusual core structure, highlighted in pink. Taking the circuit

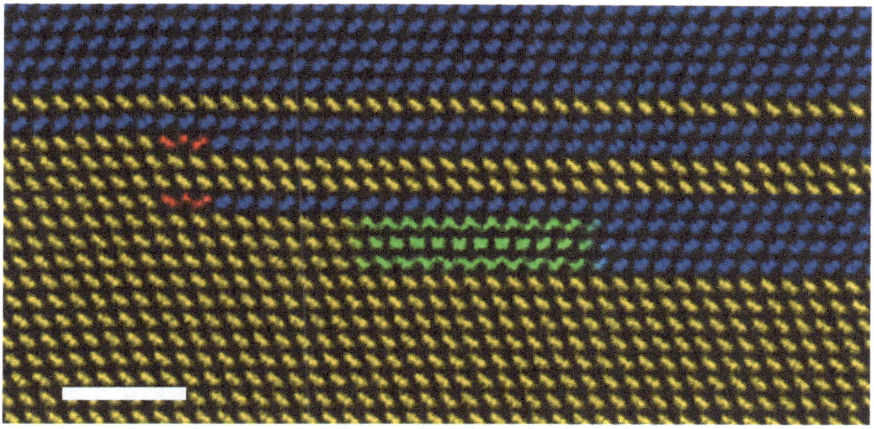

Fig. 3.7 Coloured band-pass filtered ADF-STEM image showing typical appearance of a 3ML type defect where atom columns can not be resolved and the structure has a blurred appearance. Scale bar 2 nm

$$-\mathbf{b} = 2t_1 - 5s_1 - 4t_2 + 4s_2 + s_1 = 4s_2 - 2t_1 - 4s_1 \tag{3.7}$$
$$-\mathbf{b} = \frac{4}{6}[411] - \frac{2}{2}[10\bar{1}] - \frac{3}{2}[101] - \frac{1}{2}[011] = \frac{1}{6}[11\bar{2}]$$

reveals a Shockley partial dislocation, and it is assumed the Shockley partial being next to a 3ML defect reduces the energy and makes this configuration more stable.

Through the analysis of 3ML type defects it was noticeable that the atomic columns are not always clearly resolved. Sometimes the atomic columns appear blurred over a small section of the material where the direction of the dumbbells gradually changes. An example of this is shown for a (111)-twin step in Fig. 3.7, where blurred atomic columns are highlighted in green.

In Fig. 3.7 the {112} twin boundary is not clear since the atomic configuration is not resolved, and the blurred region extends across approximately 5 nm. There are two possible explanations for their appearance. The first is the direction along which the defect is viewed. As was explored in the introduction Sect. 1.3 a ⟨110⟩ zone axis is generally chosen as the viewing direction for these defects, since the GaAs(P) dumbbell and stacking can be resolved. It was also mentioned that ZB NWs have a hexagonal shape with 6 {110} edge facets. From NW morphology, there are effectively 3 equivalent ⟨110⟩ directions available to view defect structure. If the 3ML defect extends through the entire thickness of the NW, and is viewed along one of the other two ⟨110⟩ directions, the atomic configuration will not be resolved since the defect line does not extend along the view direction. This is demonstrated in Fig. 3.8, where Fig. 3.8a shows how the defect line and twin plane change orientation if rotated by 120° with the view direction indicated by a pink arrow.

To demonstrate how the appearance of the 3ML defect changes when the view direction is changed in this way, a view of a model NW with a 3ML defect in the

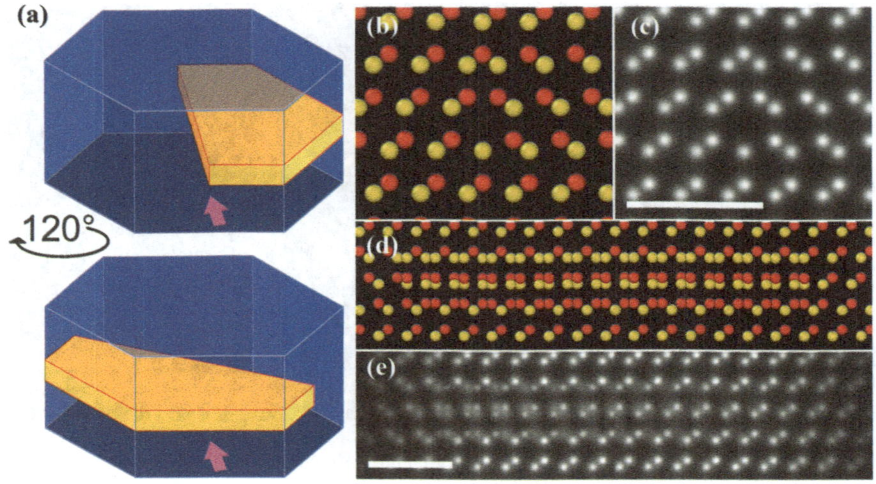

Fig. 3.8 Influence of NW orientation on a defect structure appearance. **a** In the upper NW the defect line extends in the same direction as the view direction (pink arrow), and the lower NW is rotated by 120° about the $[\bar{1}\bar{1}\bar{1}]$ direction. **b** A model 3ML defect in a NW viewed along the defect line direction, and **c** an ADF-STEM simulated image of this view. **d** A view of the rotated model and **e** an ADF-STEM simulated image of this view. Scale bars 1 nm. Model images generated using OVITO software [20]

centre (similar to Fig. 3.8a) along the defect line is shown in Fig. 3.8b. A simulated ADF-STEM image of this view is shown in Fig. 3.8c. A view of the crystal model rotated by 120° is shown in Fig. 3.8d, and a simulated ADF-STEM image of this is shown in Fig. 3.8e. When the 3ML defect is viewed in a ⟨110⟩ direction that is not the same as the defect line the dumbbells become blurred from one side of the image to the other as the dumbbell direction changes.

The other explanation for the blurred appearance is a kink at some point along the defect line. Examples of this are shown in Sect. 4.3.1 of the next chapter.

Interfacial steps or twinned segments will also have a zero Burgers vector when they are multiples of the 3ML type defect. Figure 3.9a contains a 6ML (labelled 1) and a 9ML defect (labelled 2), with both of these being of the twin variety. Figure 3.9b shows an example of five 3ML type defects in close proximity, where all are within 5 nm of the nearest 3ML defect. A final example of multiple 3ML defects is shown in Fig. 3.9c. This is a 6ML step type whose associated strain is shown in Fig. 3.9d. Here again the GPA analysis confirms no strain associated with this type of defect, since it is a zero Burgers vector defect with no long-range strain. The tallest example observed with multiple 3ML defects was a defect 60MLs tall.

Although rare, having been seen only 5 times throughout this study of defective NWs, there is another configuration of the 3ML type defect that has been observed. This configuration consists of a twinned section of material bounded by two 3ML defects, and two examples of this are shown in Fig. 3.10. If polarity in the growth

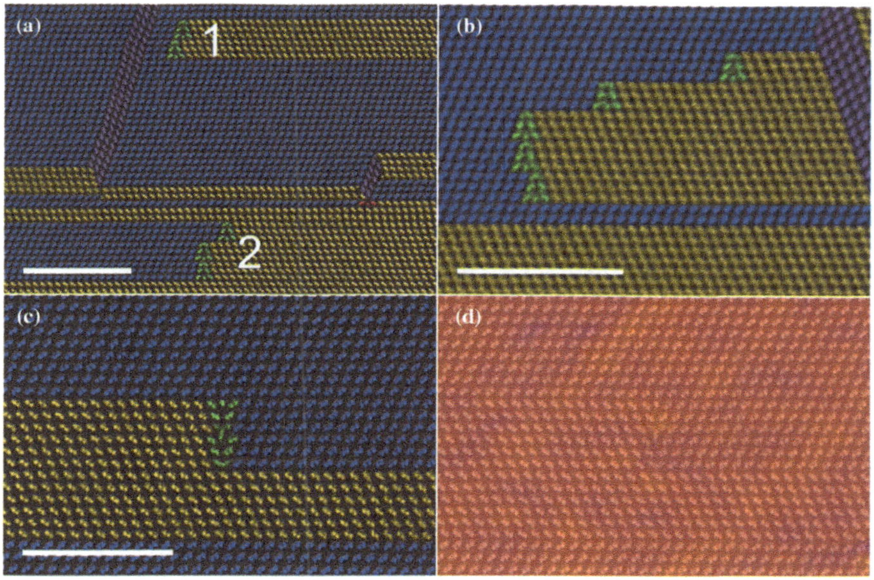

Fig. 3.9 Coloured band-pass filtered ADF-STEM images containing different configurations of multiple 3ML type defects. 3ML defects are highlighted in green. **a** An area where 1 marks an example of a 6ML twinned section type, and 2 marks a 9ML twinned section type. **b** A collection of five 3ML type defects together. **c** A 6ML step-type variant and the ϵ_{xy} strain component obtained from GPA analysis is shown in (**d**). The strain scale bar is the same as the scale bar in Fig. 3.4b. All scale bars 5 nm

direction is conserved, then the core of each bounding 3ML defect will be different [15, 18]. Figure 3.10 demonstrates that both types of 3ML defect with a III-III core and a V-V core coexist in the same structure, and the concept is demonstrated schematically in Fig. 3.10c.

So far, the 3ML defects have been observed in $(\bar{1}\bar{1}\bar{1})$ twinned areas, i.e. along the growth direction, generating a $(11\bar{2})$ para-twin. Nevertheless, a rare configuration where the 3ML type defect has been seen as an 'inclined' variety, i.e. for example the defect seen in Fig. 3.10b rotated 70° anticlockwise. This type occurs when a $(11\bar{1})$ twin ends the twinned section of material. An example of this configuration is shown in Fig. 3.11. Figure 3.11a shows a low magnification image of the defect, and Fig. 3.11b shows that the 3ML section of twinned material is stopped by a $(\bar{1}\bar{1}\bar{1})$ twin boundary. The core structure is highlighted in green in Fig. 3.11c.

Sometimes a 3ML twinned section terminates at a less common $(11\bar{1})$ twin boundary. A typical example of this is shown in Fig. 3.12. For completeness, the Burgers vector is verified by taking the vector circuit labelled in Fig. 3.12 which can be described by

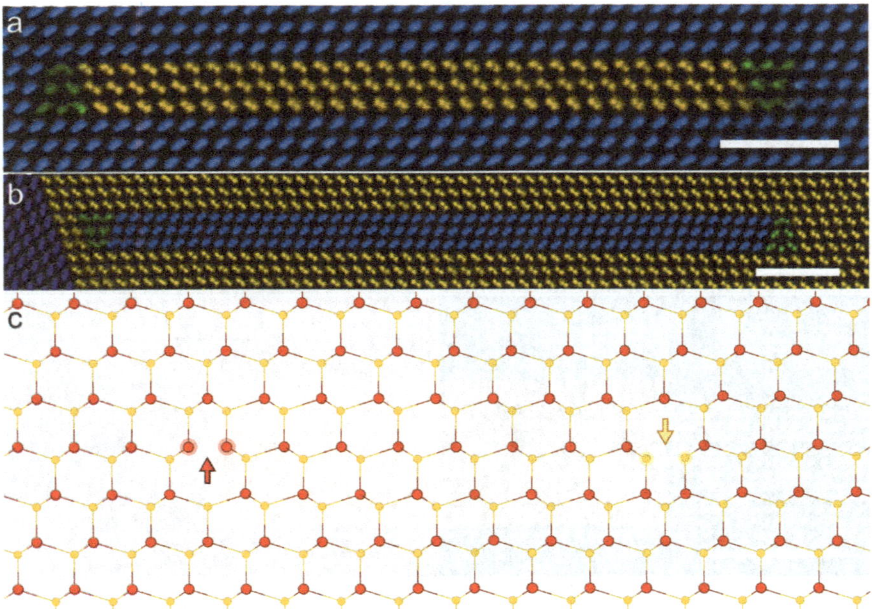

Fig. 3.10 **a**, **b** Coloured band-pass filtered ADF-STEM images of a section of twinned material bounded by two $\Sigma 3\{112\}$ twin boundaries (3ML defects). **c** A model diagram of the bonding configuration. Two arrows indicate two different core structures (III-III bond and V-V bond). Scale bars 2 nm. **a** Adapted with permission from [21]. Copyright 2019 American Chemical Society

$$-\mathbf{b} = 3t_1 - s_1 - 3s_2 - t_1 - t_3 - t_1 + 4s_1 + t_1 + t_3 = 2t_1 + 3s_1 - 3s_2 \quad (3.8)$$
$$-\mathbf{b} = 0$$

The vector sum is the same as the one used to calculate **b** for the 3ML type in Eq. 3.4. This is another example where the multiple twinning complicates Burgers vector analysis, and the use of this approach makes defect analysis easier. While the appearance differs, it looks like a 3ML twin boundary is cut in half, it still has the same **b**.

3.2.3 Defects with Non-Zero Burgers Vector

Whilst the 3ML type defect was observed to be the most common, there are number of other types of defects that have been observed. Other defects observed typically include partial dislocations, Hirth locks, LCLs and perfect dislocations. A common characteristic of these non-zero **b** defects is that they are typically found to be trapped by interactions with other defects, frequently with a zero net Burgers vector that is

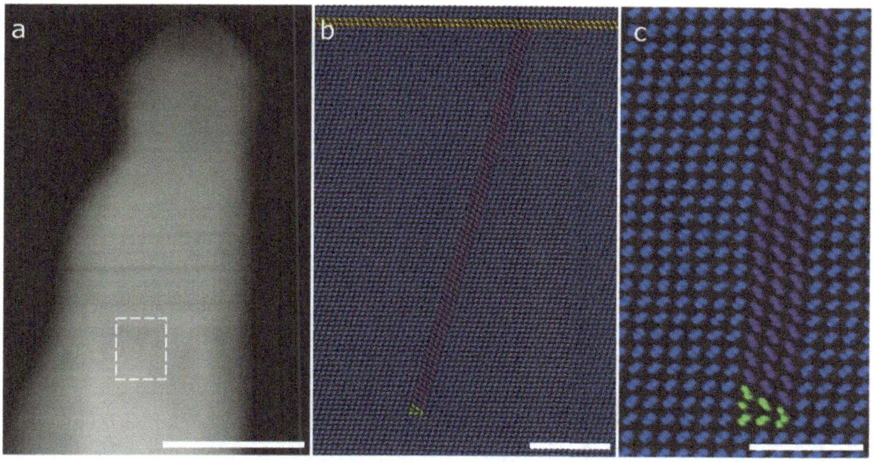

Fig. 3.11 **a** Low magnification ADF-STEM image of a defective NW tip (scale bar 100 nm). Lines of contrast indicate twinning. The area highlighted with the dashed box shows an inclined line of contrast. A higher magnification image of this area is shown in (**b**) and (**c**). **b**, **c** High magnification coloured band-pass filtered ADF-STEM images of a 3ML type defect in the uncommon $(11\bar{1})$ plane. Towards the top of (**b**) the 3ML section of twinned material is stopped by a twin in the $(\bar{1}\bar{1}\bar{1})$ plane. The green highlights the core structure. Scale bars 5 nm in (**b**) and 2 nm in (**c**)

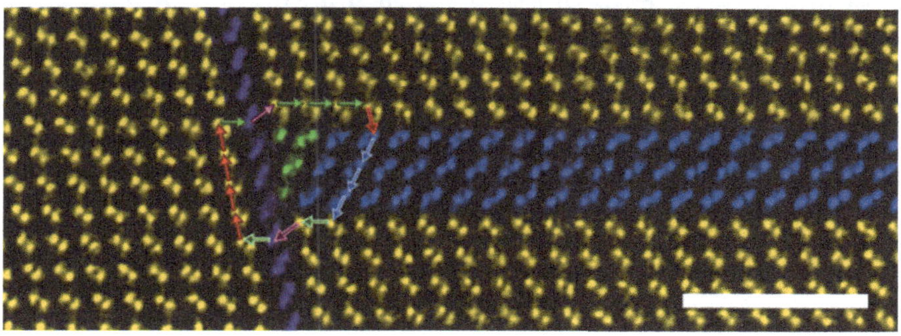

Fig. 3.12 Coloured band-pass filtered ADF-STEM image of a 3ML defect interacting with a $(11\bar{1})$ twin. A circuit is drawn around the structure to perform Burgers vector analysis. Scale bar 2 nm

likely to lock them in place. Some examples have been observed where two $\mathbf{b} \neq 0$ defects together form a $\mathbf{b} = 0$ configuration.

While many of the defects shown in this section are widely observed, some of the examples include structures that would be unusual to find in bulk material. NWs can therefore provide an opportunity to study uncommon defect structures. In the following, the observed defects with non-zero Burgers vector will be described in terms of the number of MLs they disturb.

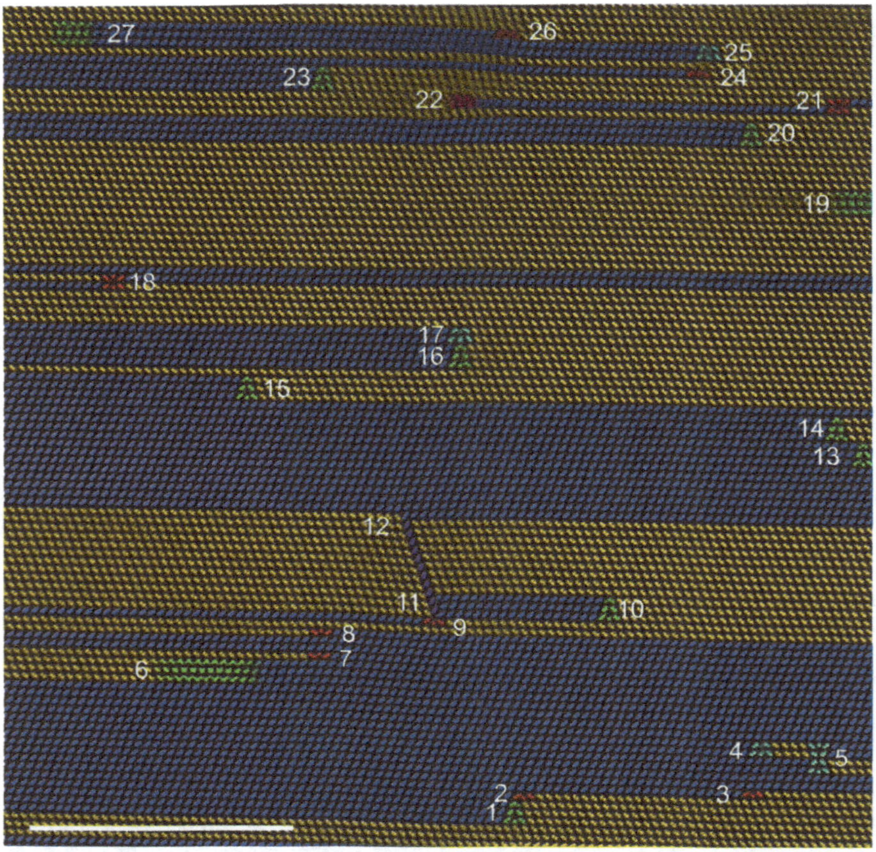

Fig. 3.13 Coloured band-pass filtered ADF-STEM image of a defective region in the NW tip that was shown in Fig. 3.1d. Defect structures are highlighted in different colours according to the type of defect. 1ML structures in red, 2ML in blue, 3ML in green. Scale bar 10 nm. Adapted with permission from [22]. Copyright 2018 American Chemical Society

To begin this section, an area roughly 30 nm by 30 nm in size that contains 27 defects has been analysed (Fig. 3.13). The first areas in this image to be examined demonstrate a few common features. Figure 3.14a corresponds to a magnified image of defects 16 and 17 from Fig. 3.13. On first inspection, it is a twinned section of material 5MLs in height terminated by a {112} twin facet. This structure can be described as the combination of a 3ML and 2ML type defect (coloured green and cyan respectively). A vector circuit around the entire structure, and starting from the top left can be described by

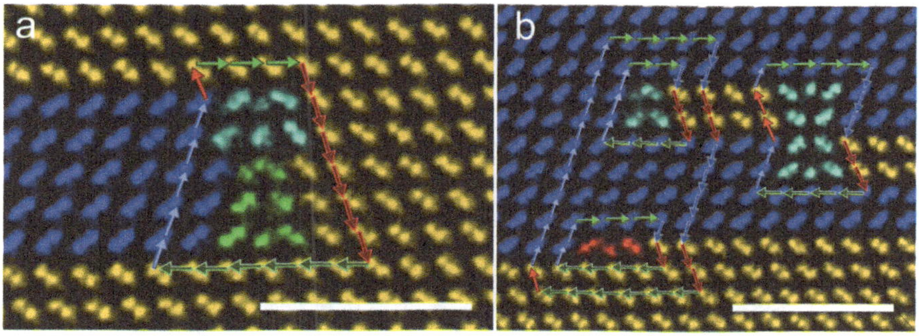

Fig. 3.14 Magnified coloured band-pass filtered ADF-STEM images of **a** defects 16 and 17 and **b** 3, 4 and 5 from Fig. 3.13. Vector circuits are drawn around each structure for Burgers vector analysis. Scale bars 2 nm. Adapted with permission from [22]. Copyright 2018 American Chemical Society

$$-\mathbf{b} = 3t_1 - 6s_1 - 6t_1 + 5s_2 + s_1 = 5s_2 - 5s_1 - 3t_1 \tag{3.9}$$
$$-\mathbf{b} = \frac{5}{6}[411] - \frac{5}{2}[101] - [10\bar{1}] - \frac{1}{2}[01\bar{1}] = \frac{1}{6}[\bar{1}2\bar{1}]$$

This gives the same Burgers vector of a single 2ML defect, i.e. an extrinsic 30° partial. Presumably, being next to a 3ML defect slightly reduces the energy of the partial dislocation and stabilises the configuration. It will be seen in Sect. 4.2 that {112} twin boundaries become more stable the taller the boundary is. Note that the core structure of the partial component here matches the lower 2 layers of the 3ML defect.

Figure 3.14b shows three structures of interest. At the bottom left of the image is a 1ML intrinsic 30° partial with the core structure coloured red. The vector circuit drawn around this defect is described by

$$-\mathbf{b} = 3t_2 - s_2 - s_1 - 4t_1 + 2s_2 = s_2 - s_1 - t_1 \tag{3.10}$$
$$-\mathbf{b} = \frac{1}{6}[411] - \frac{1}{2}[101] - \frac{1}{2}[10\bar{1}] = \frac{1}{6}[\bar{2}11]$$

which confirms an intrinsic 30° partial dislocation. At the top left of the image, there is a 2ML extrinsic 30° partial dislocation, confirmed by the circuit around it described by

$$-\mathbf{b} = 2t_2 - s_2 - 2s_1 - 3t_2 + 3s_2 = 2s_2 - 2s_1 - t_1 \tag{3.11}$$
$$-\mathbf{b} = \frac{2}{6}[411] - \frac{2}{2}[101] - \frac{1}{2}[10\bar{1}] = \frac{1}{6}[\bar{1}2\bar{1}]$$

Here the 2ML extrinsic partial dislocation (coloured cyan in Fig. 3.14b) has the same structure as the upper two MLs of a 3ML defect. If a circuit is now drawn around

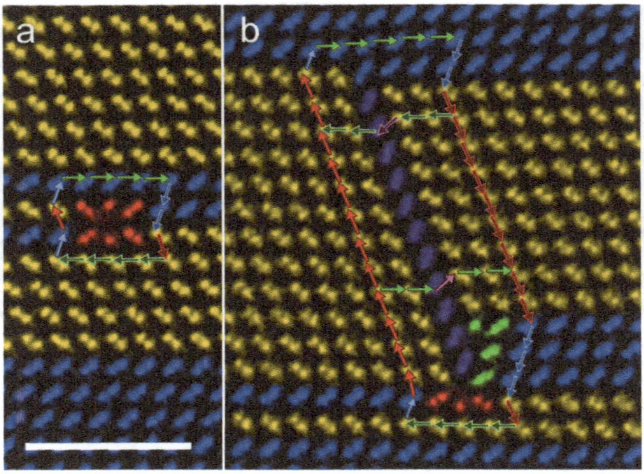

Fig. 3.15 Magnified coloured band-pass filtered ADF-STEM images of **a** defect 18 and **b** defects 9, 11 and 12 from Fig. 3.13. Vector circuits are drawn around each structure for Burgers vector analysis. Scale bar 2 nm. Adapted with permission from [22]. Copyright 2018 American Chemical Society

both of the 1ML and 2ML defects, the circuit can be described by

$$-\mathbf{b} = 4t_2 - 2s_2 - 2s_1 - 4s_2 - 2s_1 - 6t_1 + s_1 + 9s_2 = 3s_2 - 3s_1 - 2t_1 \quad (3.12)$$

which is the same as the net circuit from Eq. 3.4 and so has net $\mathbf{b} = 0$. There is an attractive force holding them in place. This configuration forms a dislocation dipole with their long-range strain fields cancelling. Another similar example of this can be seen as defects 24 and 25 in Fig. 3.13. Here, any partial dislocation seen will either have a similar partner along the same $\{111\}$ plane, extend to the surface or will be met by another type of defect structure.

The final feature of interest in Fig. 3.14b is the pair of 2ML defects on the right side of the image. This shows an example of a defect dipole where the defect is mirrored. This defect, despite being 4ML in height, can be considered to be a jog in the 2ML extrinsic stacking fault, changing its position by two layers. By simple inspection of the circuit drawn around both, all the vectors cancel each other out, and so the structure has net Burgers vector of zero. Therefore, despite the complexity of the structure observed in Fig. 3.14b containing many 'defects', if considered together the net Burgers vector is zero and no long-range strain is associated with this area.

A similar analysis was carried out in 1ML defects, with an example shown in Fig. 3.15a, which corresponds to a magnified image of defect 18 in Fig. 3.13. Here, a similar defect to the one described on the right of Fig. 3.14b with 1ML intrinsic partial dislocations is observed, with the core coloured red. The configuration can be described as a jog in the 1ML intrinsic stacking fault, and by simple inspection it can be seen that the Burgers vector circuit sum is zero. A similar example of this can be seen as defect 21 in Fig. 3.13.

Figure 3.15b shows an example of a less commonly observed $(11\bar{1})$ twin which leads to familiar defects with an unconventional appearance. This structure is split into three Burgers vector circuits indicated by vectors drawn on Fig. 3.15b. The first of these around the upper defect is described by

$$-\mathbf{b} = 5t_2 - 2s_2 - s_1 - 2t_1 - t_3 - 2t_1 + 2s_1 + s_2 = t_1 + s_1 - s_2 - t_3 \quad (3.13)$$
$$-\mathbf{b} = \frac{1}{2}[10\bar{1}] + \frac{1}{2}[101] - \frac{1}{6}[411] - \frac{1}{6}[41\bar{1}] = \frac{1}{3}[\bar{1}\bar{1}0]$$

The structure corresponds to a LCL, which is the result of the intersection of two intrinsic stacking faults on different $\{111\}$ planes [23, 24], however the Burgers vector for a LCL is $\mathbf{b} = \frac{1}{6}\langle 110\rangle$. The $\mathbf{b}$ measured from the image can be explained from a reaction between a LCL with $\mathbf{b} = \frac{1}{6}[\bar{1}\bar{1}0]$ and a perfect $\mathbf{b} = \frac{1}{2}[110]$ dislocation.

The circuit around the lower section of the structure in Fig. 3.15b can be described by

$$-\mathbf{b} = 2t_1 + t_3 + 2t_1 - 2s_1 - 3s_2 - s_1 - 4t_1 + s_2 + 4s_1 = t_3 - 2s_2 + s_1 \quad (3.14)$$
$$-\mathbf{b} = \frac{1}{6}[41\bar{1}] - \frac{2}{6}[411] + \frac{1}{2}[101] = \frac{1}{6}[110]$$

which shows it is a LCL whose appearance differs to expected appearance because a 3ML twin defect (coloured in green) is combined with the LCL, and since the 3ML component has $\mathbf{b} = 0$ the result is a LCL $\mathbf{b}$ value with a different core structure. A circuit around both of these is described by

$$-\mathbf{b} = 5t_2 - 2s_2 - 9s_1 - 3s_2 - s_1 - 4t_1 + s_2 + 12s_1 + s_2 = t_1 + 2s_1 - 3s_2 \quad (3.15)$$
$$-\mathbf{b} = \frac{1}{2}[10\bar{1}] + \frac{2}{2}[101] - \frac{3}{6}[411] = \frac{1}{2}[\bar{1}\bar{1}0]$$

and so together they are a dissociated dislocation locked in place. Despite having a non-zero $\mathbf{b}$ they are sessile and will not respond to the forces that tend to drive them out of the NW. The conventional method of finding Burgers vector could not be used here because of the twinning and demonstrates how easy the approach used here makes the analysis.

Another illustrative example of the applicability of this approach is shown in Fig. 3.16, where atom columns are not resolved perfectly. Figure 3.16 contains a Burgers circuit drawn around defect number 22 from Fig. 3.13. This circuit can be described by

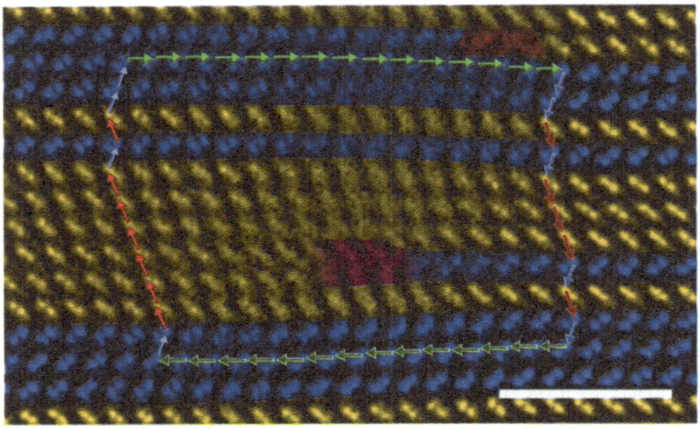

Fig. 3.16 Coloured band-pass filtered ADF-STEM image of defect 22 from Fig. 3.13. A vector circuit is drawn around the structure for Burgers vector analysis. Scale bar 2 nm. Adapted with permission from [22]. Copyright 2018 American Chemical Society

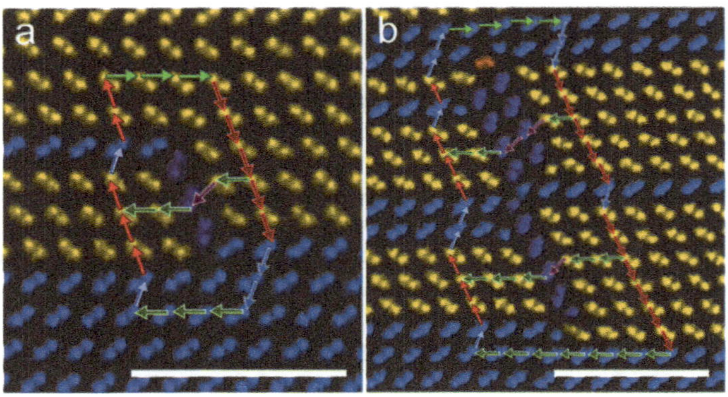

Fig. 3.17 Coloured band-pass filtered ADF-STEM images of some examples of lock type defect structures. **a** Hirth locks and **b** LCLs. Vector circuits are drawn around each structure for Burgers vector analysis. Scale bars 2 nm. Adapted with permission from [22]. Copyright 2018 American Chemical Society

$$\begin{aligned} -\mathbf{b} &= 15t_2 - 2s_2 - s_1 - s_2 - 3s_1 - s_2 - s_1 - s_2 - 14t_2 \\ &\quad + s_2 + 6s_1 + s_2 + s_1 + 2s_2 = t_2 + 2s_1 - s_2 \\ -\mathbf{b} &= \frac{1}{2}[10\bar{1}] + \frac{1}{2}[101] + \frac{1}{2}[011] - \frac{1}{6}[411] = \frac{1}{3}[111] \end{aligned} \tag{3.16}$$

revealing there is a negative Frank partial dislocation. Similar structures of this defect have been reported in CdTe [25, 26].

Figure 3.17a shows an example of a Hirth lock. The circuit around the top can be described by

$$-\mathbf{b} = 3t_1 - 3s_1 - t_1 - t_3 - 2t_1 + s_1 + s_2 + 2s_1 = s_2 - t_3 \tag{3.17}$$
$$-\mathbf{b} = \frac{1}{6}[411] - \frac{1}{6}[41\bar{1}] = \frac{1}{3}[001]$$

which shows it is a Hirth lock, with an intrinsic stacking fault on the $(11\bar{1})$ plane. The lower circuit gives

$$-\mathbf{b} = 2t_1 + t_3 + t_1 - 2s_1 - 2s_2 - 3t_2 + s_2 - 2s_1 = t_3 - s_2 \tag{3.18}$$
$$-\mathbf{b} = \frac{1}{6}[41\bar{1}] - \frac{1}{6}[411] = \frac{1}{3}[00\bar{1}]$$

which gives an equal but opposite **b** to the upper circuit. The core structures observed here are similar to those reported in CdTe [27]. The circuit around both is then

$$-\mathbf{b} = 3t_1 - 5s_2 - 2s_2 - 3t_2 + s_2 + 3s_1 + s_2 + 2s_1 = 0 \tag{3.19}$$

Which verifies that they cancel each other out and have a net Burgers vector of zero. Together these Hirth locks provide another example of a $\mathbf{b} = 0$ defect that changes the dumbbell direction in a pair of planes. Here, it occurs at the two planes that meet the extra $(11\bar{1})$ (purple) plane.

Figure 3.17b shows another example where multiple twinning contributes to the complex appearance of the defect structure. There are three distinct sections which each have twinned sections meeting and different $\{111\}$ planes intersecting. Each of these sections are analysed individually and then the complete structure is considered. A circuit drawn around the upper section can be described by

$$\begin{aligned} -\mathbf{b} &= 4t_2 - 2s_2 - 2s_1 - t_1 - 2t_3 - 2t_1 + s_1 + s_2 + s_1 + 2s_2 \\ &= t_2 + s_2 - 2t_3 \end{aligned} \tag{3.20}$$
$$-\mathbf{b} = \frac{1}{2}[10\bar{1}] + \frac{1}{6}[411] - \frac{2}{6}[41\bar{1}] = \frac{1}{6}[\bar{1}\bar{1}0]$$

which corresponds to a LCL. The appearance of this structure appears to be unusual, with an extra pair of atoms (highlighted in orange) whose dumbbell orientation does not seem to align to any of the crystal structure orientations dealt with so far (yellow, blue or purple). Within the circuit drawn here there is also a blue plane which does not terminate in a simple way with the purple plane and it takes the appearance of half a 30° intrinsic partial dislocation. The middle section can be described by

$$\begin{aligned} -\mathbf{b} &= 2t_1 + 2t_3 + t_1 - 3s_1 - s_2 - 2s_1 - 2t_1 - t_3 - 3t_1 + s_1 + 2s_2 + 2s_1 \\ &= t_3 + s_2 - 2s_1 - 2t_1 \end{aligned} \tag{3.21}$$
$$-\mathbf{b} = \frac{1}{6}[411] + \frac{1}{6}[41\bar{1}] - \frac{2}{2}[101] - \frac{1}{2}[10\bar{1}] - \frac{1}{2}[01\bar{1}] = \frac{1}{6}[\bar{1}\bar{1}0]$$

revealing another LCL, with a very different core structure. This section has a structure similar to half an extrinsic 30° partial with the two blue planes. This lock also accompanies a change in the number of extra $(11\bar{1})$ planes, leading to an uncommon and unfamiliar core structure. The lower section is described by

$$-\mathbf{b} = 3t_1 + t_3 + 2t_1 - 4s_1 - 7t_2 + s_2 + 2s_1 = t_3 + s_2 - 2s_1 - 2t_1 \tag{3.22}$$

which is the same as the middle and so $\mathbf{b} = \frac{1}{6}[\bar{1}\bar{1}0]$. This is therefore a third LCL which is also the fourth different core structure seen for a LCL. The core structure seen in the lower section has a similar appearance to previously reported examples in CdTe [27]. This core has what appears to be two intrinsic 30° partials on the (111) and $(11\bar{1})$ planes. A circuit around all three of these locks is given by

$$\begin{aligned} -\mathbf{b} &= 4t_2 - 2s_2 - 5s_1 - s_2 - 6s_1 - 7t_2 + s_2 + 3s_1 \\ &\quad + 2s_2 + 3s_1 + s_2 + s_1 + 2s_2 = 3s_2 - 4s_1 - 3t_2 \\ -\mathbf{b} &= \frac{3}{6}[411] - \frac{2}{2}[101] - \frac{2}{2}[011] - \frac{3}{2}[10\bar{1}] = \frac{1}{2}[\bar{1}\bar{1}0] \end{aligned} \tag{3.23}$$

which as expected is the sum of the **b** from the three sections and is a perfect dislocation. This is an example of a perfect dislocation that has dissociated and locked in place, due to the multiple twinning involved.

Figure 3.18a contains an area of a defective NW tip with four interesting structures close to each other. The first feature of interest (labelled as 1 in Fig. 3.18a) is a 3ML defect which is magnified with a vector circuit drawn around it in Fig. 3.18b. The core of this example is a unique variation, where the upper component has been shifted to the bottom, and the bottom layer now sits at the top. It can be described as a 1ML defect at the top and a 2ML at the bottom. The vector circuit can be described by

$$\begin{aligned} -\mathbf{b} &= 43t_1 - 4s_1 - 5t_2 + 3s_2 + s_1 = 3s_2 - 3s_1 - 2t_1 \\ -\mathbf{b} &= \frac{3}{6}[411] - \frac{3}{2}[101] - \frac{1}{2}[10\bar{1}] - \frac{1}{2}[01\bar{1}] = 0 \end{aligned} \tag{3.24}$$

which confirms **b=0** as expected from other 3ML defects. Of over 300 defects in NWs, this is the only example of this variation observed, showing it to be rare.

Figure 3.18c corresponds to a more familiar lock structure, with structure similar to that seen in the lower circuit of Fig. 3.17b. The circuit around this can be described by

$$\begin{aligned} -\mathbf{b} &= 2t_1 + t_3 - 4s_1 - 4t_2 + s_2 2s_1 = t_3 + s_2 - 2t_1 - 2s_1 \\ -\mathbf{b} &= \frac{1}{6}[41\bar{1}] + \frac{1}{6}[411] - \frac{2}{2}[10\bar{1}] - \frac{1}{2}[101] - \frac{1}{2}[011] = \frac{1}{6}[\bar{1}\bar{1}0] \end{aligned} \tag{3.25}$$

revealing a LCL. The structure towards the top of the $(11\bar{1})$ purple plane in Fig. 3.18d shows another unique core structure. Here there is what looks to be a 6ML defect

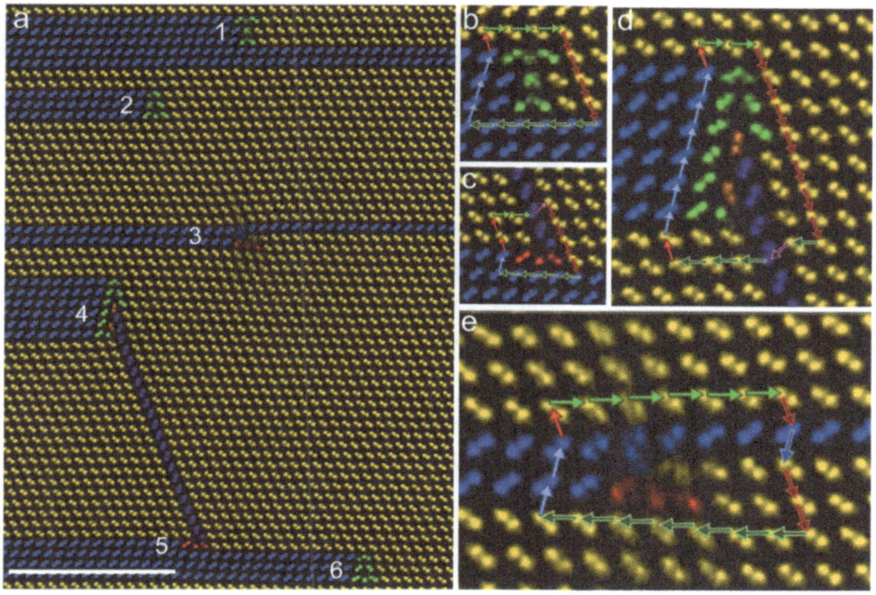

Fig. 3.18 Coloured band-pass filtered ADF-STEM images of **a** examples of Frank partial defect structures amongst others. Scale bar 5 nm. **b**–**e** An enlarged view of defects 1, 5, 4 and 3 respectively. Vector circuits are drawn around each structure for Burgers vector analysis

(highlighted in green) interacting with the $(11\bar{1})$ plane (purple). This leads to an unusual core structure, where there are 2 extra dumbbell pairs (coloured in orange) that are not orientated in any of the 3 main crystal orientations (yellow, blue or purple). A circuit is drawn around this structure and can be described by

$$-\mathbf{b} = 2t_1 - 7s_1 - t_1 - t_3 - 3t_1 + s_1 + 6s_2 + s_1 = 6s_2 - 5s_1 - 2t_1 - t_3 \quad (3.26)$$
$$-\mathbf{b} = \frac{6}{6}[411] - \frac{5}{2}[101] - \frac{1}{6}[41\bar{1}] - \frac{1}{2}[10\bar{1}] - \frac{1}{2}[01\bar{1}] = \frac{1}{3}[11\bar{1}]$$

which corresponds to the Burgers vector of a Frank partial in the uncommon $(11\bar{1})$ plane. This then shows the interaction between a Frank partial and a 6ML defect with $\mathbf{b} = 0$. The structure in Fig. 3.18e shows a more familiar positive Frank partial dislocation, and the circuit can be described by

$$-\mathbf{b} = 6t_1 - s_1 - s_2 - 2s_1 - 7t_1 + 2s_2 + s_1 = s_2 - t_1 - 2s_1 \quad (3.27)$$
$$-\mathbf{b} = \frac{1}{6}[411] - \frac{1}{2}[10\bar{1}] - \frac{1}{2}[101] - \frac{1}{2}[011] = \frac{1}{3}[\bar{1}\bar{1}\bar{1}]$$

which corresponds to a Frank partial dislocation. Similar structures of this defect have been reported in CdTe [25, 26].

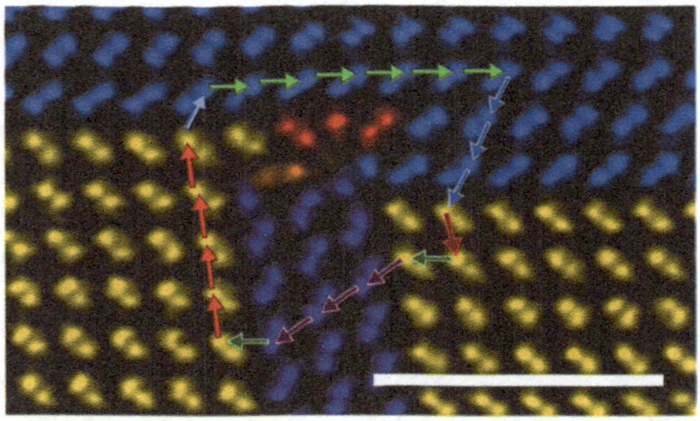

Fig. 3.19 Coloured band-pass filtered ADF-STEM image of a Frank partial dislocation with unusual core structure. A vector circuit is drawn around the structure for Burgers vector analysis. Scale bar 2 nm

An example of a Frank partial with an unusual core structure is shown in Fig. 3.19. The vector circuit drawn around this can be described by

$$-\mathbf{b} = 6t_2 - 3s_2 - s_1 - t_1 - 3t_3 - t_1 + 4s_1 + s_2 = 4t_2 - 2s_2 + 3s_1 - 3t_3 \quad (3.28)$$
$$-\mathbf{b} = \frac{3}{2}[10\bar{1}] + \frac{1}{2}[01\bar{1}] - \frac{2}{6}[411] + \frac{3}{2}[101] - \frac{3}{6}[41\bar{1}] = \frac{1}{3}[\bar{1}\bar{1}\bar{1}]$$

which corresponds to a Frank partial. The unusual appearance of this example comes from the interaction between the Frank partial and a 3ML defect. This has similarities to the 6ML interaction seen in Fig. 3.18d, the difference being this time the 3ML defect is in the purple crystal orientation.

Figure 3.20 shows an unusual example of what at first appears to be two 1ML defects in close proximity. Closer inspection shows a $(11\bar{1})$ twin extending across just two planes. This leads to an unusual appearance of a single purple dumbbell with two isolated atoms next to this in the $(11\bar{1})$ plane. The circuit around the upper section can be described by

$$-\mathbf{b} = 5t_1 - s_1 - s_2 - t_1 - t_3 - 2t_1 + 3s_1 = 2s_1 - s_2 - t_3 + 2t_1 \quad (3.29)$$
$$-\mathbf{b} = \frac{2}{2}[101] - \frac{1}{6}[411] - \frac{1}{6}[41\bar{1}] + \frac{1}{2}[10\bar{1}] + \frac{1}{2}[01\bar{1}] = \frac{1}{6}[110]$$

which corresponds to a LCL configuration. The circuit around the lower section is described by opposite vectors to the top section, and from inspection the complete circuit cancels out to give net zero Burgers vector.

This chapter has so far demonstrated a wide variety of defect structures found in NWs. These defects can largely be classified depending on the number of MLs involved as well as lock type structures. Figure 3.21 shows the relative frequency of each different type (1ML, 2ML, 3ML and lock) to determine how common each of these different type of defects are. The numbers here were obtained from a survey sampling 317 defects. These numbers show the 3ML defect to be the most commonly observed, followed by lock configurations, then 2ML, and finally 1ML defects. The

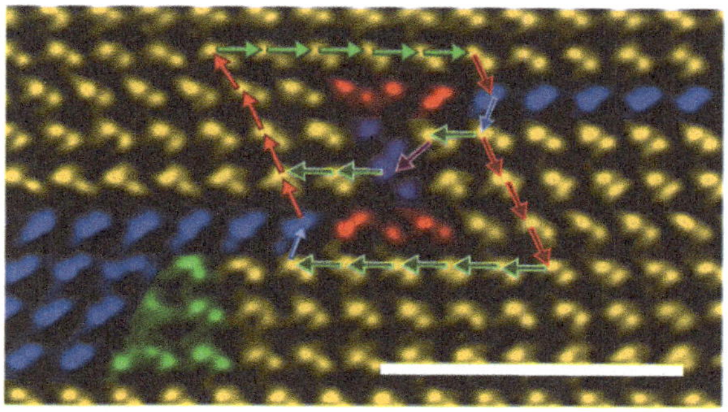

Fig. 3.20 Coloured band-pass filtered ADF-STEM image of an unusual pair of LCLs in close proximity. A vector circuit is drawn around the structure for Burgers vector analysis. Scale bar 2 nm

numbers do not distinguish between isolated 2ML or 1ML defects and those that are associated with other defects, for example those observed in dipole configurations contribute two observed defects. This shows both that the 3ML type defect is the most common, and how interactions between defects is likely to be common from the number of lock configurations seen. The most commonly seen defects are also the most stable.

In this section it has been shown how not only can a large variety of defects exist inside a NW, with a number of these recognisable and familiar, but also how these familiar defects can interact with each other and form configurations with net Burgers vector zero that lock them in place. Interactions with $\mathbf{b} = 0$ defects lead to some unique core structures that are not commonly seen in bulk or thin film materials, and the Burgers vector remains the same despite a different core structure. These unusual structures are largely a result of multiple twinning that occurs during sub-optimal growth.

In summary, the Burgers vector analysis carried out in defective NW tips shows that defects are stable inside the NWs. These defect configurations can be classified in three different groups: (i) they have a Burgers vector of zero, (ii) they are rendered immobile through locking reactions, or (iii) they are in the form of dipoles. This is quite different from the microstructure observed in bulk material and is an indication of the effect of the limited crystal volume on the types of defects that can exist. In particular, the observation of three different types of line defect with a zero Burgers vector is a new phenomenon and illustrates the uniqueness of the NW microstructure.

3.3 Defect Origin

Since NWs have a reputation for being mostly defect free, a natural question to ask after demonstrating such a wide variety of defects that can be present in a NW is how do they form? This section aims to answer this question.

To do this, a self-catalysed GaAs NW sample where the catalyst droplet has not been consumed was used. NWs were dispersed using the dropcast technique onto a

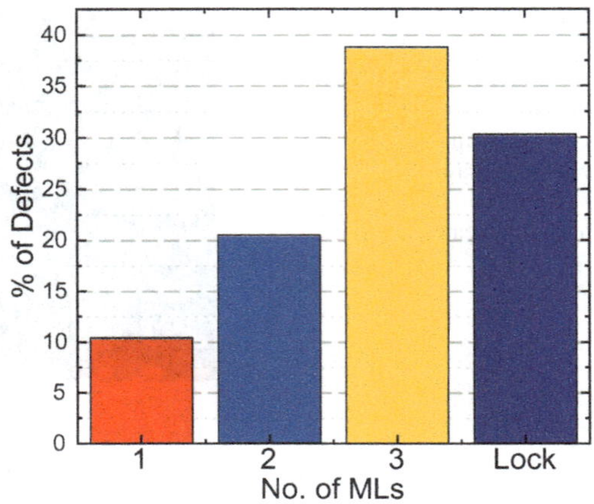

Fig. 3.21 A bar chart to show how many of each type of defects were seen from a survey of 317 defects. Adapted with permission from [22]. Copyright 2018 American Chemical Society

DENS Solutions wildfire heating chip (as described in Sect. 2.2.4). A suitable NW for viewing along a ⟨110⟩ direction with the Ga catalyst droplet still attached was chosen. A TEM image was taken of the NW inside a JEOL 2100 equipped with a Gatan Oneview CCD. The tip of the selected NW can be seen in Fig. 3.22a with a flat droplet/NW interface. A yellow line is drawn on the image at the interface to act as a reference point. Close to this interface there are a few twins, which will be used again as a reference point to see how the NW structure evolves after being heated and cooled. The dark contrast in the top half of the image is the Ga droplet, where there are also signs of an oxide shell around the droplet.

Whilst observing the NW in-situ and recording live images at 60 frames per second, the temperature of the chip was raised to 350 °C from room temperature, being careful to watch for signs the NW stem begins to decompose. A general observation noticed was the behaviour of NWs at elevated temperatures was very sensitive to the electron beam. At 350 °C more dynamic behaviour of the droplet/NW interface is seen when the electron beam is more focused on the structure. To minimise effect of electron beam, the beam condenser was not changed once temperature was raised. The temperature was then reduced by 100 °C and focus adjusted as necessary to keep the NW in focus as cooling the heating chip causes sample z height to change.

Frames during the reduction of temperature are shown in Fig. 3.22d–h, with time stamps given in the top right of the images. The full video can be found in the supporting information of [21]. Figure 3.22d is from early in the simulated growth where two distinct islands of new material can be seen, with one of these twinned. A schematic of this is shown in Fig. 3.22c where the edge of the NW tip has been approximately traced from each of the labelled frames. The blue regions on this figure indicate a twinned island of material on the left-hand side of the NW tip. Frames e–h show snapshots as time progresses, where new material builds and gradually meets. The right island of new material moves more across the interface than the left. The final two frames show the two growth fronts have made contact. On these frames,

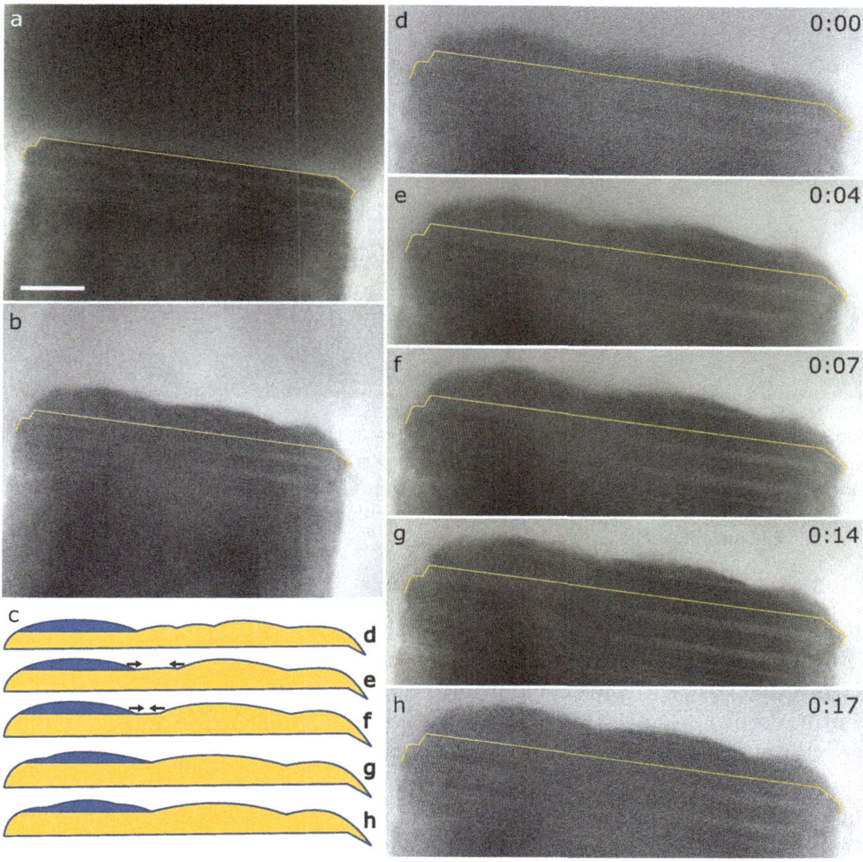

Fig. 3.22 **a** TEM image of a GaAs NW with catalyst droplet attached. The yellow line marks the NW/droplet interface. Scale bar 10 nm. **b** The same NW tip after being heated and cooled inside the TEM. **c** Outline obtained by tracing the NW edge in Figures d–h. **d**–**h** TEM frames taken at varying points in time as the NW cools and simulated NW growth proceeds. Time relative to frame d is labelled at the top right of the images. Adapted with permission from [21]. Copyright 2019 American Chemical Society

twins in the NW stem are used as a reference point to track where the original flat NW tip was, and is marked by a yellow line.

An observation to be made here is how the Ga droplet has changed after the heating and cooling process. In Fig. 3.22b, there is a noticeable difference in the Ga droplet shape. The catalyst droplet has become much smaller, and the accompanying oxide shell has reduced in size. One possible reason for the reduction in droplet size is Ga evaporation. The use of 350 °C here is below the congruent evaporation temperature (T_c) for {111}B GaAs faces which is reported to be 630 °C [28]. When GaAs evaporates into a vacuum (Langmuir evaporation), below T_c the Ga and As

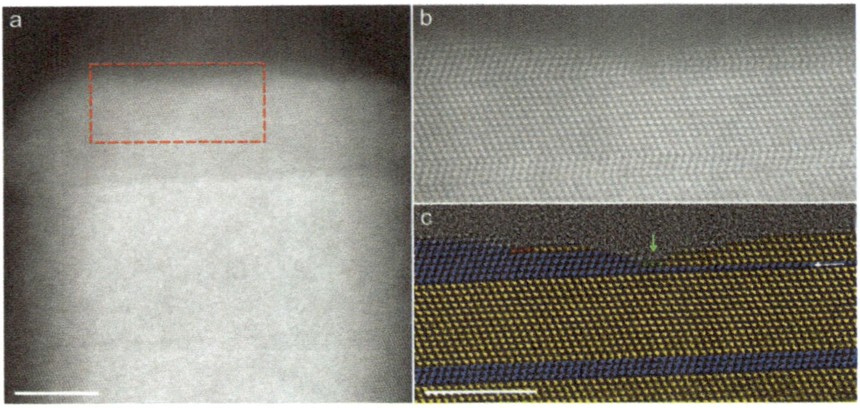

Fig. 3.23 **a** ADF-STEM image of the NW tip from Fig. 3.22 after being heated and cooled (Scale bar 10 nm). The area highlighted by the red box is shown in (**b**) at higher magnification (scale bar 5 nm). **c** A coloured and band-pass filtered version of (**b**). A green arrow indicates where two islands of new material have met. Scale bar 5 nm. Adapted with permission from [21]. Copyright 2019 American Chemical Society

fluxes leaving the surface are equal while above T_c As preferentially evaporates [29]. Another possible reason for reduction in Ga droplet size is Ga redistribution via diffusion [30]. Diffusion could include surface diffusion on NW sidewalls away from the droplet [31, 32]. Throughout this heating experiment, the oxide shell around the NW and droplet acts as a barrier to confine the material into a nano-reactor environment. This means that as the NW stem decomposes, the material enters the liquid Ga droplet inside the oxide shell. When cooled, the As that joined the catalyst droplet will form new material with Ga from the droplet. If the oxide shell is broken, the droplet is no longer restricted and material will evaporate away from the droplet into vacuum and cause the droplet to shrink and leaves no catalyst, as observed in the image series. Breaking of the oxide shell could be caused by the electron beam.

To get a clearer image of what has happened to the NW structure, the same NW was taken to an ARM200F to perform STEM imaging. ADF-STEM images of the NW tip are shown in Fig. 3.23. Figure 3.23a contains a lower magnification image, where the reduced Ga droplet size is clearly observed. The NW tip is recognisably the same, and a higher magnification image of the area highlighted by the red box is shown in Fig. 3.23b. To make analysis easier, the image is band-pass filtered and coloured (Fig. 3.23c). A stacking fault across the entire width of the NW is revealed (indicated with a white arrow), and on top of this two distinct regions of differently orientated material are present, with one blue and one yellow orientation. Where the two orientations meet (in the middle, indicated by a green arrow) is the foundations of a 3ML defect forming, highlighted in green. Interestingly, there is another twin on the left side island, which has formed a 1ML intrinsic partial (highlighted in red), or the foundation for another 3ML defect, depending on how the rest of the structure growth would continue.

While this investigation is not direct evidence of what happens inside a MBE chamber during NW droplet consumption, this does provide evidence behind the mechanism of defect formation in NWs. As was expected, it is possible for two different islands to nucleate on the tip of a NW and any twinning involved favours the formation of defects. As was seen in the introduction Sect. 1.3.1, there is a critical radius below which no defects should form. The NW from Fig. 3.22 has a radius of ≈ 26 nm, and from Sect. 1.3.1 the critical radius for GaAs is ≈ 10 nm for 30° partials and ≈ 22 nm for edge dislocations. The radius of the NW used in this example is therefore greater than the critical radius and defect formation is possible. This investigation shows that the suspected mechanism of multiple nucleation and twinning is the cause of defect formation during NW catalyst droplet consumption.

3.4 Cross-Sections of Defective Nanowires

When dealing with defect structures in NWs, an assumption made is that the defect line is straight and extends to the other side of the NW. As was seen in Sects. 3.2.2 and 3.2.3 glissile dislocations can be found in NWs, and as was seen in Sect. 1.3.2 these dislocations can move by gliding via thermally generated kink nucleation. Each kink displaces the defect by one lattice translation vector on its glide plane. The dislocations observed in Sects. 3.2.2 and 3.2.3 that can glide have {111} glide planes. With kinks being thermally generated, a defect in the process of moving at the elevated growth temperatures could be halted when the temperature is reduced towards the end of growth.

An investigation has been made to determine which of these possibilities occur in the NWs studied in this chapter. To do this, slices normal to the NW growth direction are taken from the defective tip, and examined in the [111] direction. The slicing procedure is outlined in methods Sect. 2.2.2.

A series of BF-STEM images along the [111] direction of NW cross-sections are shown in Fig. 3.24. The images are presented roughly in order of increasing structure complexity, with the NW cross-section in Fig. 3.24a showing what seems to be four sections separated by four lines of contrast, while Fig. 3.24h shows numerous lines and areas of contrast. Typical contrast features observed are shown schematically in Fig. 3.24i and include (i) a straight line extending across the entire width of NW, (ii) a kinked line and (iii) a sector of contrast.

This series of cross-section images shows how the defective tip crystal structure is perhaps even more complicated than first thought. The appearance differs from that typical of microtome knife damage. While multiple twinning is common in these NWs, the lines and contrast in these images suggest there may be multiple sections of material meeting and interacting within the NW structure, possibly from multiple island nucleations during growth. As a reminder, these NWs were first grown with a GaAsP stem, the catalyst droplet was then consumed, and finally a GaAsP shell was grown (see Sect. 2.2.1). The shell is grown via the VS mechanism, and so the structure of the NW tip will either be determined by the droplet consumption crystallisation

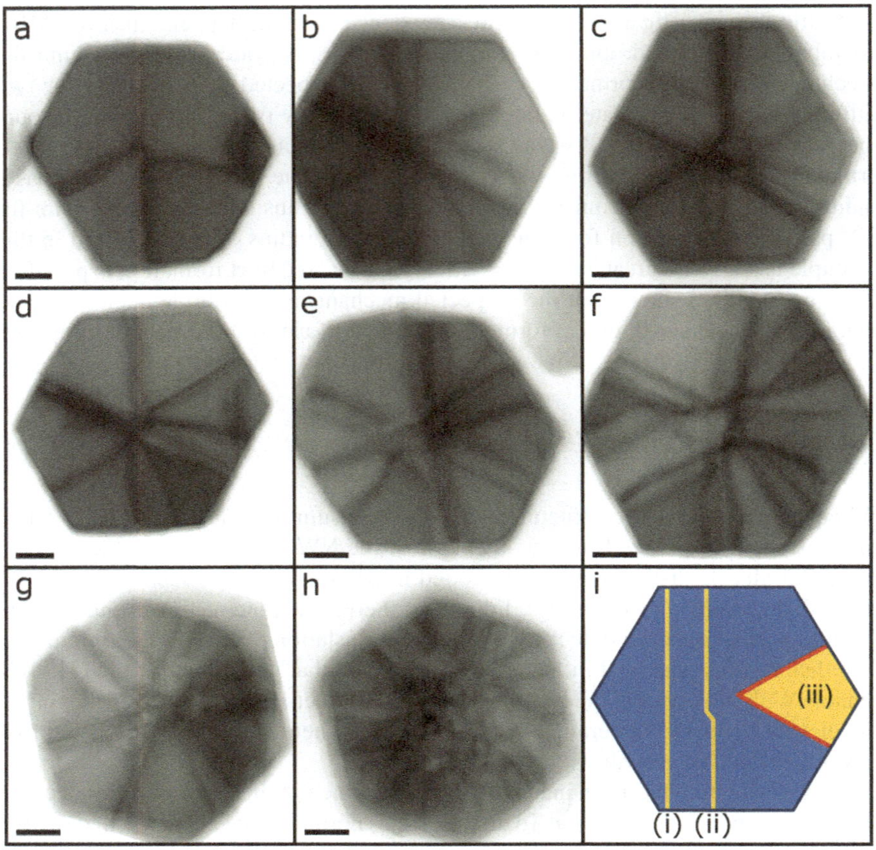

Fig. 3.24 **a–h** A collection of BF-STEM images of sliced NW cross-sections from the tip region of defective NWs. Lines of contrast indicate presence of defects. Images are presented in rough order of increasingly complicated structure. Scale bars 20 nm. **i** A schematic of typical contrast features seen in STEM images of defective NW cross-sections. (i) indicates a line of contrast that extends across the entire width of the NW, (ii) indicates a kinked line and (iii) a sector of contrast

and subsequent epitaxial radial growth from the consumed droplet, or pure axial and radial VS growth. VS growth is known to result in comparatively more stacking faults than vapour-liquid-solid (VLS) [33]. No NW cores are seen in Fig. 3.24 and so epitaxial radial growth from the NW core does not occur in this region of NW, and the precise growth mechanism of the material in the sliced region is not known.

The contrast in these images may be explained by a few possibilities. One of these is from strain of dislocations, including partial dislocations. To see evidence of this, the NW in Fig. 3.24g was tilted approximately 35.3° to a $\langle 110\rangle$ axis. An ADF-STEM image of this NW viewed along the $\langle 111\rangle$ direction and the tilted view are shown in Fig. 3.25a, b respectively. On both of these images a red box highlights a feature of

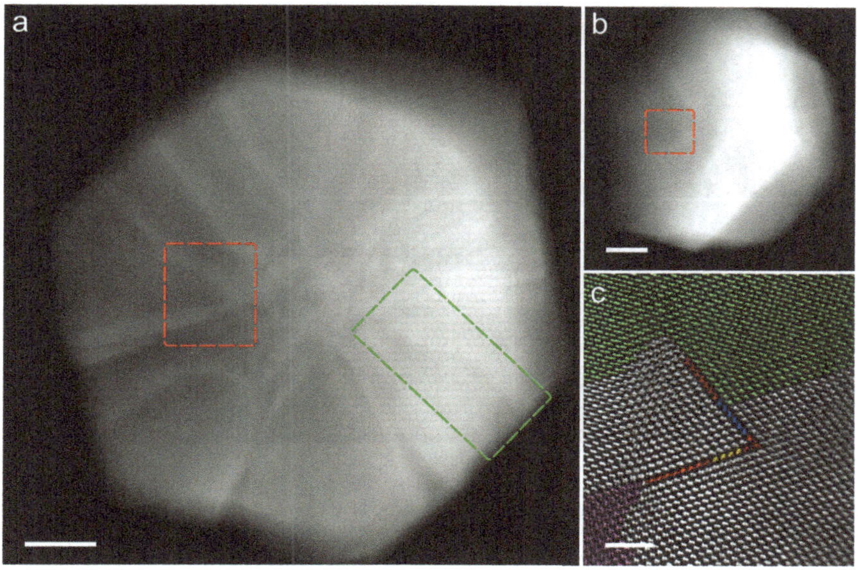

Fig. 3.25 **a** ADF-STEM image of NW cross-section from Fig. 3.24g. Scale bar 20 nm. **b** ADF-STEM image of the same NW cross-section tilted ≈ 35.3°. Scale bar 20 nm. **c** High magnification band-pass filtered ADF image of the area marked by the red box in (**b**), viewed along a ⟨110⟩ direction. The same contrast feature is highlighted by red boxes in (**a**) and (**b**). Red colouring highlights 1ML type defects while the blue and yellow indicate stacking faults in different {111} planes. Scale bar 2 nm

interest, where two lines of contrast meet towards the centre of the NW, an example of type (iii) from Fig. 3.24i. A band-pass filtered image of this feature along the ⟨110⟩ direction is shown in Fig. 3.25c and reveals a LCL with an appearance similar to previous examples seen in Figs. 3.17b, 3.18c, and 3.20. The two {111} planes involved also each have a 30° intrinsic partial, coloured in red. The atoms coloured in blue and yellow indicate stacking faults. The presence of partial dislocations like this can result in strain and produce contrast features in the image.

The image in Fig. 3.25c gives an example of another commonly seen feature in the NW slices, where the ⟨110⟩ dumbbell projection is not resolved for all areas of the image. The region towards the bottom left of the image (roughly highlighted in pink) becomes blurred and atom columns are not well resolved, and towards the top region of the image (roughly highlighted in green) extra peaks in image intensity appear and dumbbells become sets of three intensity peaks. The change in appearance may be a result of an overlap of two sections of twinned material in the direction of the electron beam where the twin boundaries are not in the direction of the zone axis and causes extra peaks of intensity in the STEM image. Overlap of two sections of twinned materials is demonstrated in Fig. 3.26. Figure 3.26a shows a side view of a NW model with a 3ML defect where the structure has been rotated

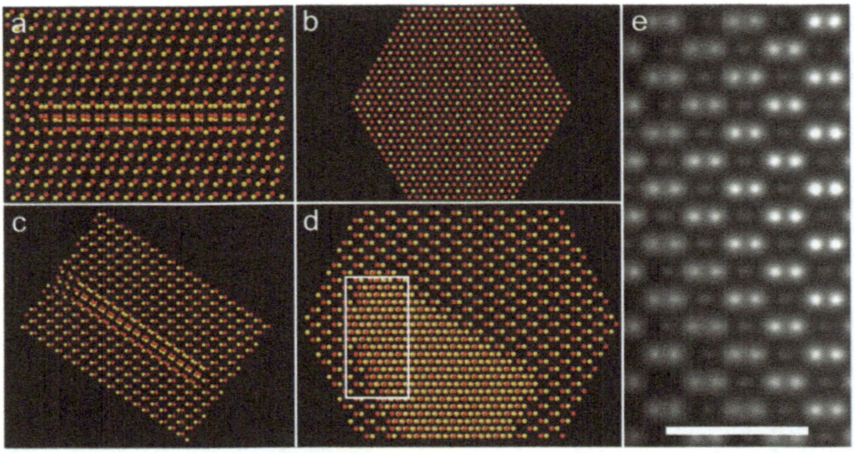

Fig. 3.26 **a** Model NW with a 3ML defect, rotated by 120°, with similar appearance to that seen in Fig. 3.8d. **b** Top down view of the structure. **c** The same NW rotated by ≈ 35.3°. **d** Top down view of the tilted structure. **e** A simulated ADF-STEM image of the region indicated by a white box in (**d**). Scale bar 1 nm. Model images generated using OVITO software [20]

by 120°, similar to Fig. 3.8. Figure 3.26b shows the top-down view in the ⟨111⟩ direction. This NW structure is tilted approximately 35.3° (Fig. 3.26c) in a similar way to the NW cross-section from Fig. 3.25, and a top-down view in a ⟨110⟩ direction is presented in Fig. 3.26d. The top-down view in Fig. 3.26d demonstrates how the typical appearance of dumbbells in the ⟨110⟩ direction is changed by overlap of twinned structures, and a simulated ADF-STEM image of the region indicated by the white box in Fig. 3.26d is shown in Fig. 3.26e. The simulated ADF-STEM image has additional peaks of intensity between dumbbells, similar to experimental image observations in Fig. 3.25c. A similar change in atomic resolution images of a ZB material viewed in a ⟨110⟩ direction when the material has inclined {111} twins was reported in GaSb NWs by Zamani et al. [34].

The area marked by a green box in Fig. 3.25a shows what is a suspected example of a kinked line defect, with the line of contrast changing direction in the marked region, an example of type (ii) from Fig. 3.24i. Two more typical examples of contrast features are highlighted in Fig. 3.27. The red box in Fig. 3.27a marks what is suspected to be a defect which extends through the entire thickness of the NW in a single straight line, an example of type (i) from Fig. 3.24i. Any line of contrast is almost always along a ⟨110⟩ direction, which could correspond to many of the defect types seen earlier in the chapter that have a defect line in a ⟨110⟩ direction. The yellow box in Fig. 3.27b shows another band of contrast in the BF image which stops suddenly towards the NW centre. It was seen in Fig. 3.11 that inclined twins can occur in the NWs studied. The overlap of twinned sections of material having different crystal orientations can produce contrast in the image, and so bands of contrast like the area marked in Fig. 3.27b could be explained by the presence of inclined twins.

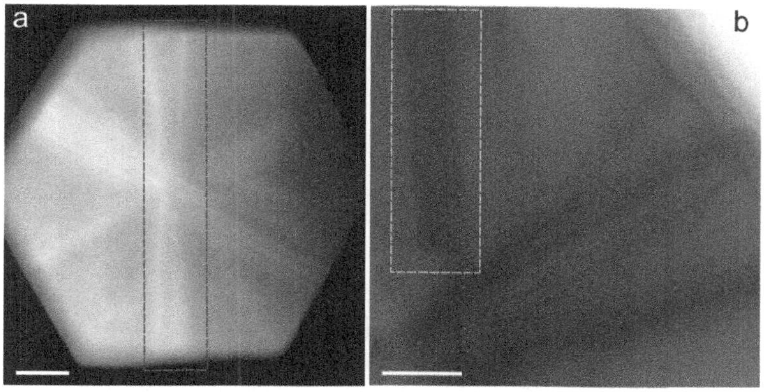

Fig. 3.27 **a** ADF-STEM image of a defective NW tip cross-section from Fig. 3.24b. The red box shows where a line of contrast extends across the entire width of the NW. Scale bar 20 nm. **b** A higher magnification BF-STEM image of a region of NW cross-section from Fig. 3.24e. The yellow box shows an area with a band of contrast which stops towards the centre of the NW. Scale bar 10 nm

Evidence of inclined twins could also suggest the presence of defects like many of those observed in Sects. 3.2.2 and 3.2.3.

To bring this section to a close, it has been shown that the defective region in NW tips may be even more complicated than first thought, with signs of multiple twinning in planes perpendicular to growth direction. This may highlight how the depth of focus of STEM will be important when interpreting atomically resolved images of NW structures since there may be structural features deeper inside the NW. Some explanations for contrast in STEM images of defective NW cross-sections have been given, although further work would be required to reach more concrete conclusions.

3.5 Effect on Nanowire Properties

To measure the suspected detrimental effect these defects have on optical properties of the NW, CL was performed on NWs with defective tips. An example of an NW with defective tip is shown in Fig. 3.28a. A panchromatic CL image is shown in Fig. 3.28b and a combination of the SE SEM image and panchromatic CL image is shown in Fig. 3.28c. Clearly there is no emission from the defective tip region and so non-radiative recombination centres severely reduces emission from this region. There is also a noticeable variation in CL intensity in the panchromatic CL image. This is likely to be related to the P content variation in the NW. NWs from this sample have shown some P content variation along the length of the NW in EDX scans, and an example EDX linescan is shown in Fig. 3.29a where only the P variation is shown. The inset of Fig. 3.29a shows an ADF-STEM image of the NW this measurement

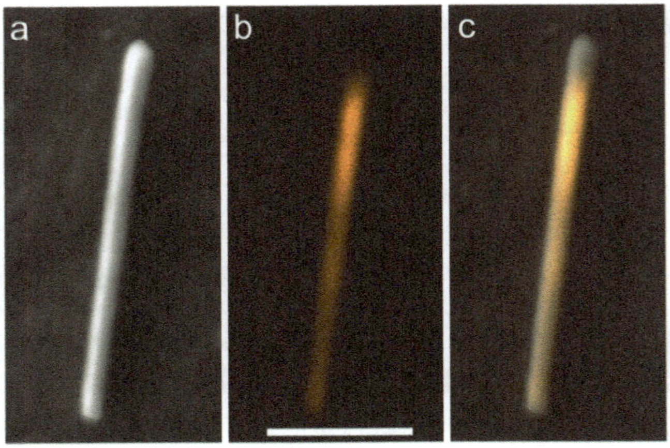

Fig. 3.28 **a** SE SEM image of a NW with a defective tip. Accelerating voltage 5 kV. **b** Coloured panchromatic CL map image of the NW with stage at $-100\,^{\circ}$C. **c** Panchromatic CL image overlaid with the SE SEM image, showing no emission from the tip region of the NW. Scale bar 1 μm. Adapted with permission from [22]. Copyright 2018 American Chemical Society

was taken from and the line indicates where the EDX linescan was performed. This shows the P content varies along the NW, and the lower P content towards the tip of the NW will be from the shell section of NW growth which was grown with a lower P content.

To compliment this, point CL spectra were acquired from two positions along the NW, one from the middle region and one from the tip region. These spectra are both shown in Fig. 3.29b. The spectrum from the mid region shows a broad peak around 680 nm, which corresponds to emission from $GaAs_{0.67}P_{0.33}$ according to Eq. 1.2 in Sect. 1.2.3. There is another peak visible around 630 nm, which is attributed to background noise. This background peak is still seen in the tip spectrum and acts as a good reference point to confirm that the now absent 680 nm peak shows emission is quenched in the defective tip region.

This has been taken a step further, and another NW was first examined by taking a CL map and then imaged at high magnification with ADF-STEM imaging. A SE SEM image with a panchromatic CL map image overlaid is shown in Fig. 3.30a. An ADF-STEM image of the same NW is shown in Fig. 3.30b and confirms the same NW is being imaged. The overlaid CL map in Fig. 3.30a shows no emission from the tip region of the NW. A higher magnification BF-STEM image of the same NW tip is shown in Fig. 3.30c. The contrast here indicates the presence of defects in this NW tip. Atomically resolved ADF-STEM images of two chosen areas are shown in Fig. 3.30d, e. These images reveal $\Sigma3\{112\}$ twin boundaries (coloured green), extrinsic 30° partial dislocations (cyan) and a 4ML defect (pink). This provides evidence that the types of defects examined throughout this chapter are responsible for the quenching of emission in the tip region of GaAsP NWs whose catalyst droplet has been consumed.

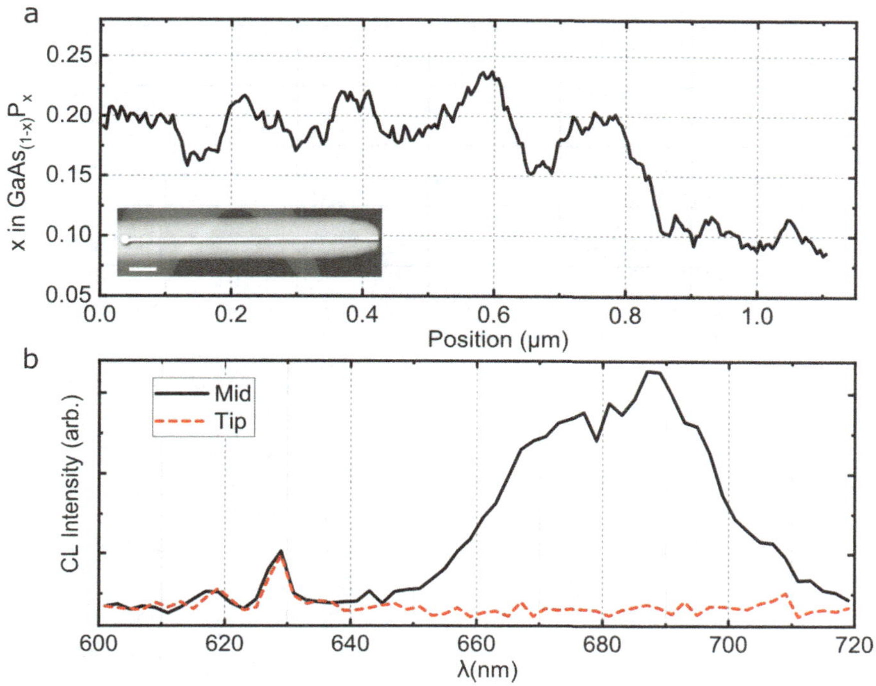

Fig. 3.29 **a** Smoothed EDX linescan taken along the line indicated on the inset ADF-STEM image of the NW (scale bar 100 nm). **b** CL spectra taken from points along the middle (solid black line) and tip region (dashed red line) of the NW shown in Fig. 3.28. **b** Adapted with permission from [22]. Copyright 2018 American Chemical Society

3.6 Chapter Summary

This chapter has acted as a showcase for how extensive and varied the possible type of defect structures inside NWs can be. Defect structures varied from the relatively simple ortho-twin interfaces and stacking faults to partial dislocations interacting with uncommon twin planes locking them in place. The frequency of multiple twinning has shown configurations of well known defects that differ from their usual appearance as seen in bulk and thin films. NWs have therefore proven to act as an interesting system to study defect structures with some novel defect configurations. The $\Sigma 3\{112\}$ twin boundary, in this chapter referred to as the 3ML defect, has been found to be the most common defect in the NWs studied. The abundance of this defect is detrimental to NW performance, since they act as non-radiative recombination centres [18], with evidence of optical emission quenching seen by CL analysis performed on NWs with defective tips.

In-situ annealing experiments to simulate NW growth inside a TEM provided information about the origin of these defects. It was shown that for large enough

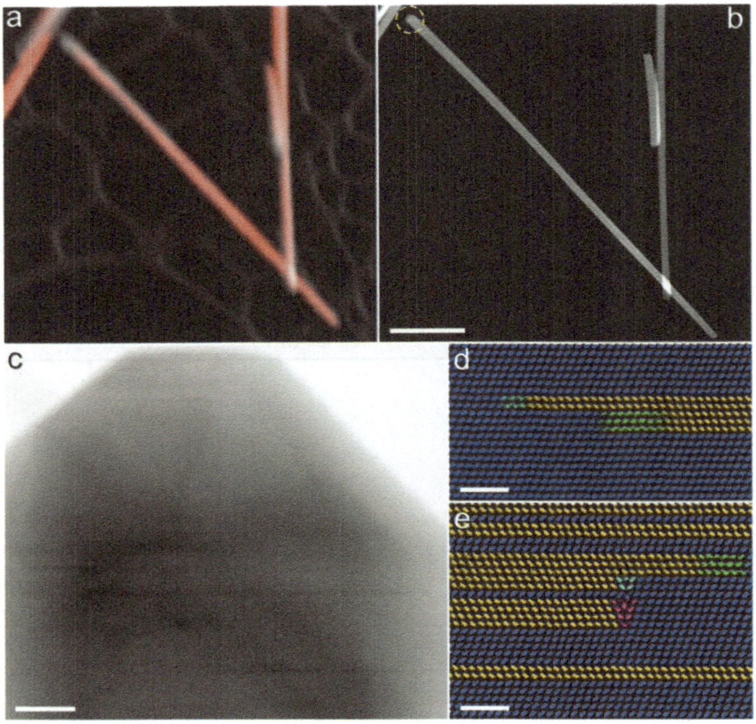

Fig. 3.30 **a** SE SEM image of a NW with a CL map image overlaid in red. There is no emission seen from the tip region of the NW. **b** ADF-STEM image of the same NW with the tip region marked by a dashed yellow circle (scale bar 1 μm). **c** BF-STEM image of the tip region of the NW (Scale bar 20 nm). **d**, **e** Coloured band filtered ADF-STEM images of two areas in the tip region of the NW. (Scale bars 2 nm)

radius (roughly > 20 nm), multiple islands of new material can nucleate, and if some of these islands are twinned, it can lead to the different defect structures that have been observed.

From analysis of NW cross-sections, it has been concluded that the behaviour of the defects through the thickness of the NW may not be as simple as first thought. Evidence consistent with defect lines kinking through the structure and likely interactions between defects deep inside the NW may disturb the defect line from simply extending from one side of the NW to the other.

References

1. Johansson J et al (2006) Structural properties of ⟨111⟩B-oriented III-V nanowires. Nat Mater 5:574–580

2. Karlsson LS et al (2007) Understanding the 3D structure of GaAs ⟨111⟩B nanowires. Nanotechnology 18:485717
3. Caroff P, Bolinsson J, Johansson J (2011) Crystal phases in III-V nanowires: from random toward engineered polytypism. IEEE J Sel Top Quantum Electron 17:829–846
4. Zhang Y, Wu J, Aagesen M, Liu H (2015) III-V nanowires and nanowire optoelectronic devices. J Phys D Appl Phys 48:463001
5. Zamani RR, Arbiol J (2019) Understanding semiconductor nanostructures via advanced electron microscopy and spectroscopy. Nanotechnology 30:262001
6. Xu L et al (2008) Structure and migration of (112) step on (111) twin boundaries in nanocrystalline copper. J Appl Phys 104:113717
7. Lu N, Du K, Lu L, Ye HQ (2014) Motion of 1/3 ⟨111⟩ dislocations on Σ3{112} twin boundaries in nanotwinned copper. J Appl Phys 115:024310
8. Marquis EA, Hamilton JC, Medlin DL, Léonard F (2004) Finite-size effects on the structure of grain boundaries. Phys Rev Lett 93:156101
9. Tanaka K, Kohyama M (2002) Experimental and theoretical study of $\Sigma = 3$ incoherent twin boundary in β-SiC. Philos Mag A 82:215–229
10. Sawada H, Ichinose H, Kohyama M (2006) Gap states due to stretched bonds at the (112) Σ3 boundary in diamond. J Phys Condens Matter 19:026223
11. Paxton AT, Sutton AP (1988) A simple theoretical approach to grain boundaries in silicon. J Phys C Solid State Phys 21:L481–L488
12. Kohyama M (2002) Computational studies of grain boundaries in covalent materials. Model Simul Mater Sci Eng 10:R31–R59
13. Sakaguchi N, Ichinose H, Watanabe S (2007) Atomic structure of faceted Σ3 CSL grain boundary in silicon: HRTEM and ab-initio calculation. Mater Trans 48:2585–2589
14. Feng CB, Nie JL, Zu XT, Al-Jassim MM, Yan Y (2009) Structure and effects of vacancies in Σ3 (112) grain boundaries in Si. J Appl Phys 106:113506
15. Cohen D, Carter CB (2003) $\Sigma = 3$, {112} lateral twin boundaries in GaP. Interface Sci 11:391–401
16. Feng C et al (2012) Possible effects of oxygen in Te-rich Σ3 (112) grain boundaries in CdTe. Solid State Commun 152:1744–1747
17. Liu C-Y et al (2016) Self-passivation rule and structure of CdTe Σ3 (112) grain boundaries. Phys Rev B 93:205426
18. Sanchez AM et al (2017) Nonradiative step facets in semiconductor nanowires. Nano Lett 17:2454–2459
19. Zamani M et al (2018) Optimizing the yield of A-polar GaAs nanowires to achieve defect-free zinc blende structure and enhanced optical functionality. Nanoscale 17080–17091
20. Stukowski A (2010) Visualization and analysis of atomistic simulation data with OVITO-the open visualization tool. Model Simul Mater Sci Eng 18
21. Gott JA et al (2019) Defect dynamics in self-catalyzed III–V semiconductor nanowires. Nano Lett 19:4574–4580
22. Sanchez AM et al (2018) Stable defects in semiconductor nanowires. Nano Lett 18:3081–3087
23. Hirth J, Lothe J (1982) Theory of dislocations. Krieger Publishing Company
24. Nabarro F, Duesbery M (2002) Dislocations in solids. Elsevier Science
25. Li C et al (2013) From atomic structure to photovoltaic properties in CdTe solar cells. Ultramicroscopy 134:113–125
26. Dasilva YAR, Kozak R, Erni R, Rossell MD (2017) Structural defects in cubic semiconductors characterized by aberration-corrected scanning transmission electron microscopy. Ultramicroscopy 176:11–22
27. Paulauskas T et al (2014) Atomic scale study of polar Lomer-Cottrell and Hirth lock dislocation cores in CdTe. Acta Crystallogr Sect A 70:524–531
28. Goldstein B, Szostak DJ, Ban VS (1976) Langmuir evaporation from the (100), (111A), and (111B) faces of GaAs. Surf Sci 57:733–740
29. Heyn C, Jesson DE (2015) Congruent evaporation temperature of molecular beam epitaxy grown GaAs (001) determined by local droplet etching. Appl Phys Lett 107:161601

30. Fauske VT et al (2016) In situ heat-induced replacement of GaAs nanowires by Au. Nano Lett 16:3051–3057
31. Pankoke V, Sakong S, Kratzer P (2012) Role of sidewall diffusion in GaAs nanowire growth: a first-principles study. Phys Rev B 86:085425
32. Krogstrup P et al (2013) Advances in the theory of III–V nanowire growth dynamics. J Phys D Appl Phys 46:313001
33. Güniat L, Caroff P, Fontcuberta i Morral A (2019) Vapor phase growth of semiconductor nanowires: key developments and open questions. Chem Rev 119:8958–8971
34. Zamani RR, Gorji Ghalamestani S, Niu J, Sköld N, Dick KA (2017) Polarity and growth directions in Sn-seeded GaSb nanowires. Nanoscale 9:3159–3168

Chapter 4
Defect Dynamics in Nanowires

4.1 Introduction

As demonstrated in Chap. 3, a high density of line defects can exist in self-catalysed III-V NWs due to the rough interfaces formed during the catalyst droplet consumption stage of growth. The origin of these defects is from the instability of the growth front between the crystal NW and the liquid Ga droplet as it is consumed, when the Ga flux is switched off [1, 2]. Roughening of the interface can produce multiple islands and defects can form when they meet. Evidence for the formation of these defects was seen in the previous chapter. Once defects like this form, any subsequent epitaxial shell growth will reproduce defect structures present in the core. Defects will be detrimental to optoelectronic properties and device performance.

The structures observed in Chap. 3 can be broadly categorised into two types: (i) defects with $\mathbf{b} = 0$ formed at twin boundaries and (ii) crystal dislocations that are locked-in by reaction and/or dissociation. The most commonly observed defect was a type (i) defect, the $\Sigma3\{112\}$ twin boundary (3ML defect), which has previously been seen in GaAsP NWs [3], GaAs NWs [4] and in Cu nanocrystals [5].

Recent studies of the SiC system have looked into dynamics of partial dislocations using molecular dynamics simulations [6, 7], where it was shown that motion of partial dislocations with a non-zero Burgers vector is expected. Sarikov et al. showed the 3ML defect configuration to be stable [7].

This chapter focuses on how a defective NW system responds to being exposed to elevated temperatures. The effect of exposure to high temperatures on a defective NW system is of interest since it could be a simple way to remove defects from NWs that are detrimental to NW performance. The responses from a selection of defect types shown previously in Chap. 3 are presented. Here, selected NWs were systematically exposed to high temperatures for a controlled period of time and images were taken after each exposure. Examples where high temperatures (>500 °C) were applied and the samples imaged live using STEM are also shown. The system and defects are analysed based on defect configuration and compared to expected behaviour/stability

J. A. Gott, *Defects in Self-Catalysed III-V Nanowires*, Springer Theses,
https://doi.org/10.1007/978-3-030-94062-1_4

by considering the forces that act on them. It is shown that the common 3ML type defect with a Burgers vector of zero can move in the NW system provided temperature is sufficiently high.

4.2 Forces Behind Defect Motion in Nanowires

In conventional systems, only defects with a non-zero Burgers vector are expected to move if they are close to a surface, an applied shear stress on the slip plane exerts a force on the dislocation line, or by interaction of strain fields with other dislocations in close proximity. While this is also expected in NWs for defects with long-range strain field, i.e. with a non-zero Burgers vector, for the 3ML defects some considerations must be taken into account. Like a dissociated crystal dislocation, the attractive forces between the constituent partials prevent their independent movement. However, unlike a dislocation, their long-range strain fields cancel, and to first order a 3ML defect is neither affected by stress fields nor generates a surface image force.

Line defects such as dislocations and 3ML defects increase the Gibbs free energy of a NW. The increase in energy depends on parameters such as the length of the defect and, for dislocations, the strain field induced in the crystal, which are a function of the location of the defect. If a defect moving lowers the Gibbs free energy, then there is an energy gradient present and it can be considered to be a force acting on the line defect. The net force acting on any given defect depends on its exact configuration.

This section will focus on configurations of the 3ML type defect, and a schematic of configurations is shown in Fig. 4.1. Figure 4.1a shows a three dimensional perspective view of a NW with some 3ML defects where the {112} twin interfaces extend through the NW structure. The short purple arrow indicates the [110] direction. Figure 4.1b corresponds to the appearance of different configurations as viewed along the [110] direction. For clarity in this chapter, the NW is divided into two regions, (i) NW thickness changes with x and (ii) NW thickness is constant. Configuration 1 in Fig. 4.1 is a 3ML defect in region (i) where a twinned section of material extends to the left side of the NW. Configuration 2 in Fig. 4.1 is another 3ML defect in region (i) but the twinned section of material extends to the right side of the NW. Configuration 3 is a 3ML defect in region (ii) that extends to the left side of the NW. Configuration 4 in Fig. 4.1b is a defect that involves two 3ML defects in region (i). Towards the top of Fig. 4.1b is a 3ML step in region (ii).

To illustrate how forces can change based on configuration, schematic examples of 3ML defects are shown in Fig. 4.2. For defects like the 3ML which consist of a section of twinned material, there are multiple twin facets associated with the defect. A schematic of this in a hexagonal NW is shown in Fig. 4.2a, where a {111} twin boundary (purple) and a {112} twin boundary (green) are labelled. The energy per unit area associated to these twins is given by γ_{111} and γ_{112}. The movement of the

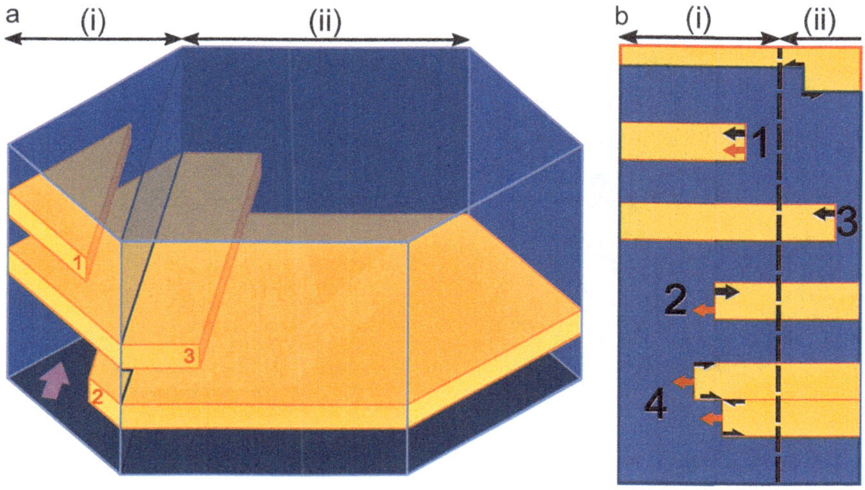

Fig. 4.1 **a** Configurations of 3ML thick twins in a NW. The purple arrow indicates the [110] direction. **b** The appearance of the different configurations in (**a**) when viewed along the [110] direction. The vertical dashed line indicates the boundary between regions (i) and (ii), with a tapering thickness and parallel sides respectively. Adapted with permission from [8]. Copyright 2019 American Chemical Society

3ML defect will either destroy or create two {111} twin boundaries. Since a twin boundary has an associated energy per unit area γ_{111}, there will be a change in energy of the system.

If a twinned section of material which ends in region (ii), or configuration 3, (Fig. 4.2c) were to move in the left direction (Fig. 4.2d) then there is a reduction in the area of the {111} twin, the purple area labelled δA. This has an associated change in energy

$$\delta E_{111} = 2\gamma_{111}\delta A \tag{4.1}$$

The factor of 2 comes from the twinned section of material having an upper and lower {111} twin.

If the defect line resides in region (i), or configuration 1 (Fig. 4.2b), then the length of the line defect must also be considered in the energy changes. Taking the x-direction to be to the right, if the defect boundary located in region (i) (Fig. 4.2b, e) moves by $-\delta x$ (Fig. 4.2f), then there is a similar δA of {111} twins (as in Fig. 4.2f), and also a change in the length of defect line (δL in Fig. 4.2f), and therefore in the {112} twin area. The change in energy associated to the movement of a 3ML defect located in region (i) is

$$\delta E = \delta E_{111} + \delta E_{112} = 2\gamma_{111}\delta A + \gamma_{112}H\delta L \tag{4.2}$$

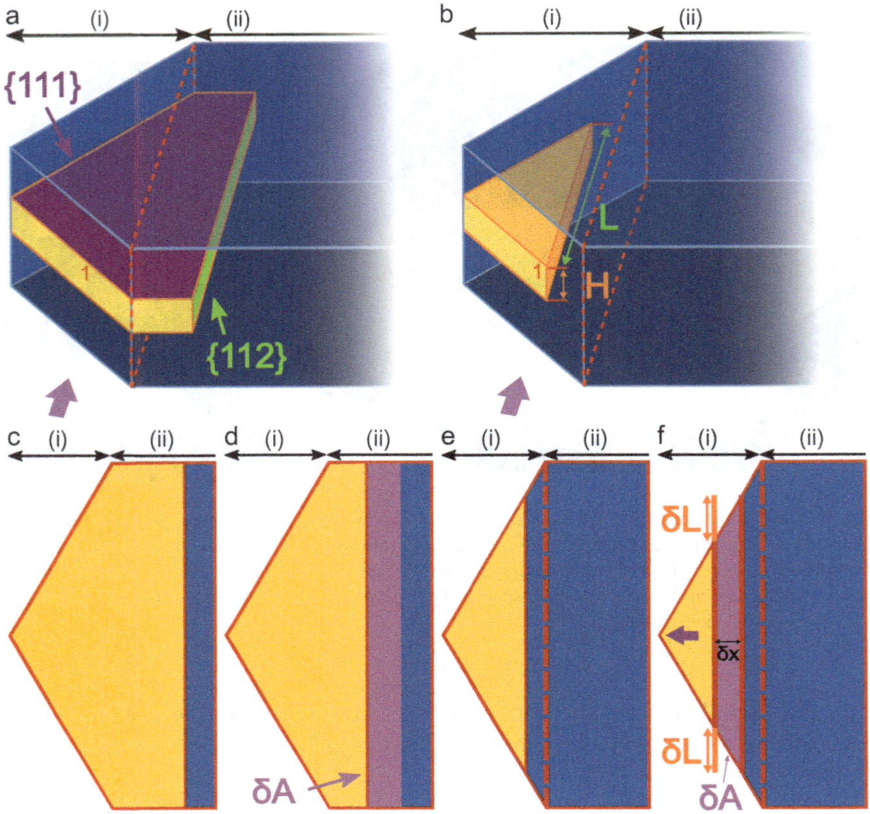

Fig. 4.2 **a** Schematic of a 3ML defect where a twinned section of material extends from the edge of the NW to the central, uniform thickness region (ii) of the NW. A {111} twin boundary is coloured purple and the {112} twin boundary is coloured green. The thick purple arrow indicates the [110] direction. **b** Schematic of a 3ML defect where a twinned section of material extends from the edge of the NW to region (i) of the NW with varying thickness. The height and length of defect are labelled by H and L respectively. **c** A top down view of the defect in (**a**). **d** The defect in (**c**) is moved towards the left, and the change in area δA of {111} twin is highlighted in purple. **e** A top down view of the defect in (**b**). **f** The defect in (**e**) is moved towards the left by δx which has a reduction in area δA of {111} twin and also a reduction in length of defect δL

where H is the height of the {112} twin boundary as labelled in Fig. 4.2b. The changes in geometry are described by

$$\delta L = -\frac{2}{\sqrt{3}}\delta x \tag{4.3}$$

$$\delta A = -L\delta x \tag{4.4}$$

Equations 4.1–4.4 provide the equations required to describe forces behind defect motion in NWs.

For configuration 1, movement of the defect to the left destroys upper and lower {111} twin boundaries and the {112} facet length shrinks. Configuration 1 moving to the left results in a change of energy

$$\delta E_{(1)} = 2\gamma_{111}\delta A + \gamma_{112} H \delta L \tag{4.5}$$

Using the geometry given by Eqs. 4.3 and 4.4, the force acting on the defect is given by

$$F_{(1)} = \frac{\delta E}{\delta x} = F_{111} + F_{112} = -2\gamma_{111} L - \frac{2}{\sqrt{3}}\gamma_{112} H \tag{4.6}$$

In this case, the negative sign means that both forces point towards the left edge of the NW. Since the height H of the defect does not change with position, the force F_{112} is independent of position in region (i) of the NW. In contrast, F_{111} depends on the length L of the defect and decreases to zero as the defect approaches the edge of the NW.

For configuration 2, movement in the $+x$ direction increases the length L of the defect but decreases the {111} twin boundary area, and the forces oppose each other

$$F_{(2)} = 2\gamma_{111} L - \frac{2}{\sqrt{3}}\gamma_{112} H \tag{4.7}$$

For configuration 3, the length of the 3ML does not change in region (ii) of the NW, and the force is

$$F_{(3)} = 2\gamma_{111} L \tag{4.8}$$

which is independent of position unless it leaves region (ii). F_{111} forces point in the direction that reduces the area of {111} twins, while F_{112} always points towards the edge of the NW, independent of the direction of {111} twinning. In Fig. 4.1b F_{111} is represented by black arrows while F_{112} is represented by red arrows. This is an influential factor in how stable defects are expected to be in NWs.

For defect configurations involving multiple 3ML defects, taking configuration 4 as an example, then in addition to the forces described by Eq. 4.7, energy is lowered by elimination of the central {111} twin boundary between the two 3ML defects. There is effectively a force acting on each 3ML defect in the direction of the other 3ML defect, indicated by half-arrows in Fig. 4.1b, and this force is described by

$$F_{(4)} = \gamma_{111} L \tag{4.9}$$

For a 3ML type defect acting as an interfacial step in region (ii) (e.g. top of Fig. 4.1b), lateral movement of the defect creates a new {111} twin boundary on one side of the defect, while removing a {111} twin boundary on the other side, and so if the length

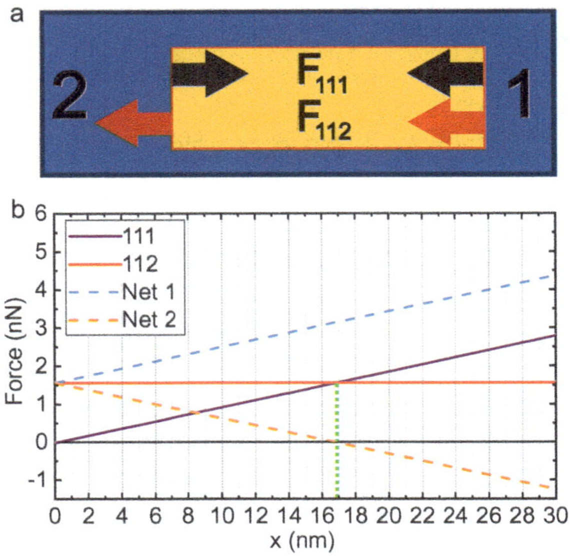

Fig. 4.3 **a** Combination of defect configurations 1 and 2. The black arrows indicate direction of F_{111} forces and the red arrows indicate directions of F_{112} forces. **b** A plot of forces F_{111} and F_{112} acting on a 3ML type defect as a function of position x along NW. The dashed line plots show the net result of forces acting on a configuration 1 and 2 type defect. The green dashed line indicates where the net force of configuration 2 is zero. Adapted with permission from [8]. Copyright 2019 American Chemical Society

of the defect does not change, there is no force acting on the defect. In this case, the defect is stable and will not move.

A combination of two defect configurations (1 and 2, Fig. 4.3a) will now be used for a demonstration of calculating the forces acting on 3ML defects in NWs. In this configuration, the net force acting on the right defect interface is expected to be larger than that of the left. In order to get some numbers for this, values for γ_{111} and γ_{112} are needed. The energy per unit area of a $\{111\}$ ortho-twin boundary is commonly estimated to be half that of an intrinsic $\frac{1}{6}\langle 112\rangle$ stacking fault [9], i.e. $\gamma_{111} \approx 27$ mJ m^{-1} in GaAs [10]. In other words, the force exerted on a 3ML defect by a twin boundary is roughly equivalent to that acting on a partial dislocation due to a shear stress of 1.2 GPa, or a strained epitaxial layer with misfit of 0.5%.

The disruption to tetrahedral bonding in the $\{112\}$ twin boundary means that γ_{112} is much larger than γ_{111}. Based on DFT modelling performed as an extension to the work done by Sanchez et al. [3], it is estimated that $\gamma_{112} \approx 980$ mJ m^{-2}. In a thick NW $L \gg H$ and the $\{111\}$ term in Eqs. 4.6 and 4.7 are expected to dominate and the opposite is true when close to NW edge in region (i). Forces acting on a 3ML defect as a function of position in region (i) of a NW are shown in Fig. 4.3b. Plots include F_{111}, F_{112}, and the net forces for configurations 1 and 2, where $|F_{(1)}| > |F_{(2)}|$. With F_{112} varying with NW thickness for configuration 2 there will be a position in the NW where $F_{(2)} = 0$. The point where $F_{(2)} = 0$ is marked in Fig. 4.3b by a vertical green dashed line and the position is estimated to be approximately 17 nm. The numbers shown here are calculated using the two γ_{112} and γ_{111} values along with height $H = 0.979$ nm given by three times the height of a ML, $h = \frac{a}{\sqrt{3}}$ [11], using the lattice constant a for GaAs from Table 1.2. At this position the 3ML defect is

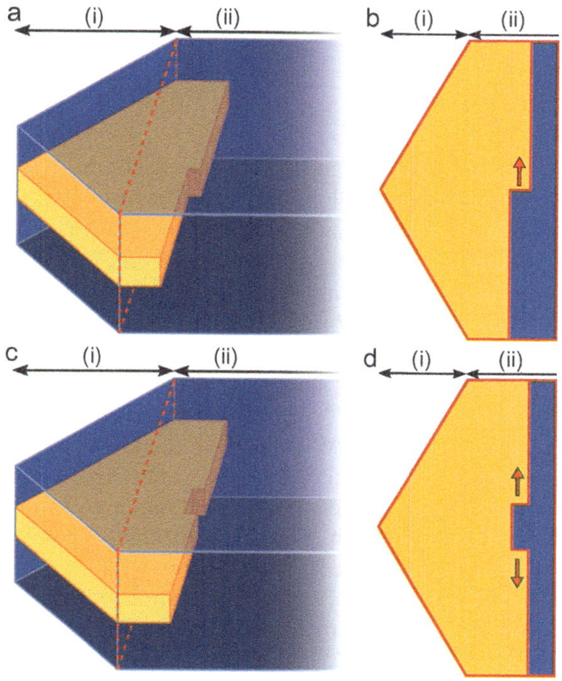

Fig. 4.4 How a line defect motion progresses in a NW via kink nucleated glide. **a, b** show single kink nucleation while **c, d** show double kink nucleation

expected to be stable and not move. In the case of twin segments which are thicker than 3ML (like configuration 4 in Fig. 4.1b), $F_{(112)}$ is larger and shifts the point of stability deeper into the NW. The concepts being discussed here all assume the NW is a perfect hexagonal shape.

The arguments presented here show that most 3ML defects are only thermodynamically stable under limited circumstances in a NW. To understand why they are seen, their mechanism of movement must be considered. Describing the 3ML defect as being composed of an intrinsic-extrinsic pair of $\frac{1}{6}\langle 112\rangle$ partial dislocations with opposite Burgers vectors, then the well-established Peierls model of dislocations [12] may be used to describe motion. In bulk material, dislocation motion is controlled by the nucleation of kink-pairs and diffusion of the kinks along the core [12]. Figure 4.4 shows illustrative examples of defect motion in a NW via single-kink nucleation (Fig. 4.4a, b) and double-kink nucleation (Fig. 4.4c, d). For long dislocations, the velocity is controlled by the kink-pair nucleation rate per unit length of dislocation; kinks travel along the dislocation until they collide with a kink of opposite sign and annihilate. For short dislocations, like the 3ML defects in these NWs, kinks are more likely to reach the surface before meeting a kink of opposite sign. In this case, the probability of kink-pair nucleation, and so velocity, is proportional to the length of the defect. The velocity of a 3ML defect is then expected to be described by the relation [13]

$$v = cLF \exp\left(-\frac{Q}{k_B T}\right) \tag{4.10}$$

$$= v_0(x) \exp\left(-\frac{Q}{k_B T}\right) \tag{4.11}$$

where c is a constant, and Q is an average of the activation energies for kink-pair formation and diffusion along both constituent $\frac{1}{6}\langle 112\rangle$ partial dislocations. The prefactor v_0 in Eq. 4.11 combines the constants from Eq. 4.10 and is a function of position in x. Equations 4.6–4.8 show that the force F depends at least in part on the defect length L for all configurations. If motion is limited by double-kink nucleation (Fig. 4.4c, d), v_0 should vary $\propto L^2$, and a slowing of a 3ML defect as it approaches the edge of the NW might be expected. However, single-kink nucleation (Fig. 4.4a, b) where the defect reaches the NW surface may also occur, which would give $v_0 \propto L$. Movement of a 3ML defect in a NW may be expected to vary considerably depending on its exact location in a NW of varying thickness. If height of the {112} twin facet is considered, a 3ML defect requires coordinated movement of both its constituent partial dislocations, while a 6ML defect requires coordinated movement of four partial dislocations, and so on. Velocities are therefore expected to decrease with increasing height of the {112} facet. The activation energy Q of these defects should be similar to that of a dissociated crystal dislocation, which is also composed of partial dislocations. A typical estimate of activation energy for $\frac{1}{2}\langle 110\rangle$ edge dislocation glide in GaAs is 1.3 eV [14, 15].

In this section it has been shown how most null Burgers vector defects are expected to be unstable in a NW but have a kinetically limited (thermally activated) motion that means they will not move out of a NW without a high-temperature anneal. Different behaviour is expected in an anneal depending on the exact defect configuration, though in general it is expected they should shrink in size until they disappear. Larger twin facets should move more slowly. While isolated cases of these defects will be expected to move as described here, there are likely to be cases of configurations that prevent straightforward movement and defects may remain stable and locked in place. The complexity of some NW structures seen in Chap. 3 provides examples where defects being locked in place may be expected.

In the following sections, in-situ heating TEM experiments are used to examine how a selection of defect structures in NWs respond to high temperatures.

4.3 Motion of 3 Monolayer Defects

The theoretical considerations made about the motion of 3ML defect in the previous section (4.2) assumes a perfect hexagonal prism NW. Figure 4.5a corresponds to a top-down view in-lens SEM image of the NWs on Si substrate used in this study. This image shows that while the majority of the NWs have the hexagonal shape, there are some NWs with a different shape (indicated by dashed circles in the

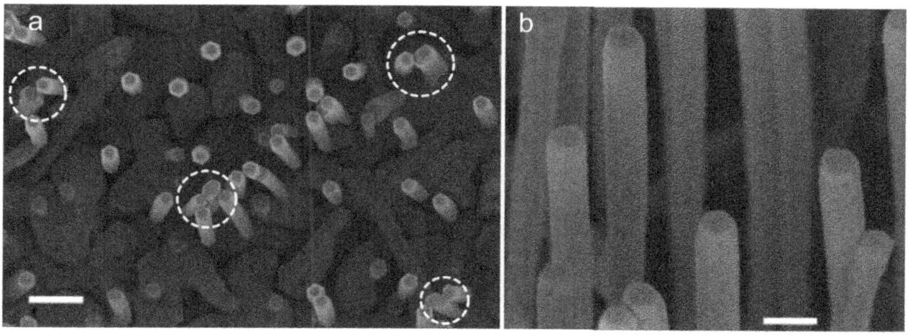

Fig. 4.5 In-lens SEM image top-down view of NWs with defective tips still attached to the substrate. Accelerating voltage 1 kV. Scale bar 500 nm. **b** In-lens SEM image of some tips of NWs attached to the substrate. Accelerating voltage 5 kV. Scale bar 200 nm. Adapted with permission from [8]. Copyright 2019 American Chemical Society

figure). These non-hexagonal NWs may show a different defect motion behaviour. Figure 4.5b shows a side view in-lens SEM image of some NW tips. This image shows the NW surface at the defective tip region is not completely flat and smooth, which may also influence defect behaviour.

To study effect of elevated temperatures on the defects, NWs were transferred to a DENS solutions Wildfire heating chip as described in methods Sect. 2.2.4. Two approaches have been used to analyse defect dynamics in NWs using in-situ heating chips. One approach involves recording live STEM images with a relatively fast scan speed while the temperature is elevated. An example of this will be presented later in the chapter. The main approach involved taking STEM images of some selected areas of some NWs, raising the temperature of the chip using a programmed heat cycle, then lowering the temperature and imaging the same areas previously selected. The heating chips typically took around 5 s to change temperature by around 250 °C. During heating and cooling the sample was observed continuously at low magnification, typically with image size 512 × 512 pixels and a dwell time of 10 μs/pixel. After each heating cycle the sample was cooled immediately to 400 °C to examine the structure at high magnification. Each heat cycle used a chosen period of time and so when tracking any changes in position of defects, the velocity can be obtained. The temperatures and times of exposure used are given in Table 4.1. A plot of temperature against time is shown in Fig. 4.6 and illustrates the heating process used for the first 6 heat cycles of Table 4.1. A total of three NWs were observed in this way over seven heating cycles and one NW was observed for a further five cycles. BF-STEM images of the three NWs chosen to perform the study are shown in Fig. 4.7. NW1 and NW3 show a more tapered shape than NW2, with NW2 having approximately vertical sides, while NW1 and NW3 have sides angled at approximately 10° from the vertical.

After exposure to heating cycles, three types of behaviour were observed for 3ML defects and include (a) complete removal from the NW (b) initial movement,

Table 4.1 Temperatures and times used for in-situ STEM observation of defective GaAsP NWs

Cycle No.	Temperature (°C) (±1°C)	Time (s)	Total Exposure Time (s)
1	600	30	30
2	620	30	60
3	620	30	90
4	620	30	120
5	640	30	150
6	640	30	180
7	680	30	210
8	680	30	240
9	700	30	270
10	700	20	290
11	700	20	310
12	700	20	330
13	700	20	350

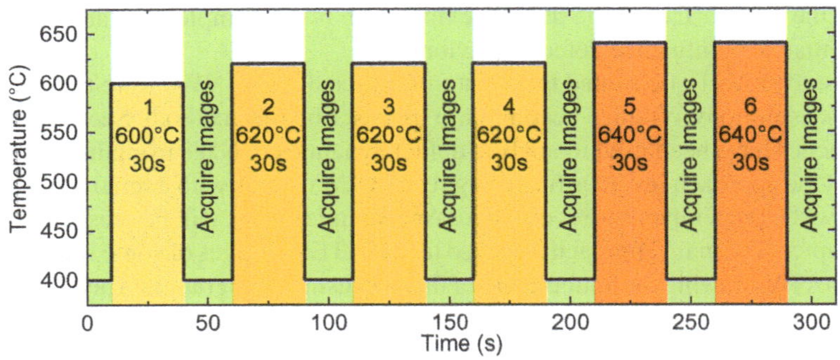

Fig. 4.6 An illustrative plot of how temperature was changed and NWs imaged to track defect motion. Cycles 1 to 6 from Table 4.1 are shown, where the temperature was raised to the labelled value for 30 s, temperature lowered to 400 °C while images were acquired, and the temperature was increased for the next heat cycle

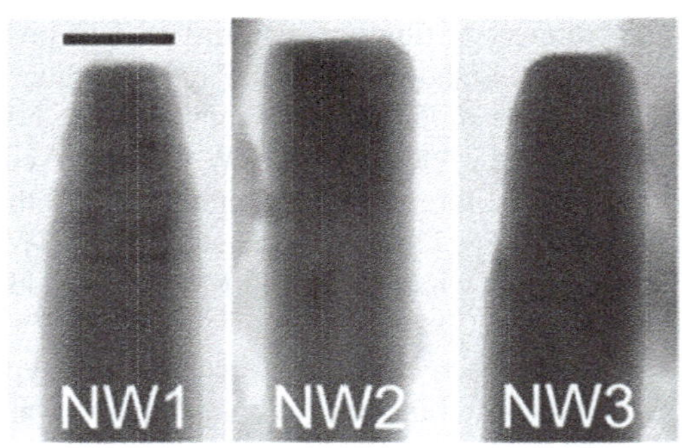

Fig. 4.7 BF-STEM images of three NWs on a heating chip used to study defect dynamics. Lines of contrast indicate presence of defects. Scale bar 100 nm. Adapted with permission from [8]. Copyright 2019 American Chemical Society

followed by a halt in a stable configuration (e.g. trapped by other defects), and (c) no movement meaning the defect is already in a stable configuration. Each of these categories will be examined in the subsequent sections.

4.3.1 Complete Removal of Defect from the Nanowire

The first example evaluates the theoretical forces described in the previous section in a real system, and corresponds to a complete removal of a defect from the NW system. The defect is a combination of configurations 1 and 2 and was originally shown in the previous chapter (Fig. 3.10), and an image from before any heat cycles are applied is shown in Fig. 4.8a.

Here there is a twinned section of material with two $\Sigma3\{112\}$ twin boundaries roughly 15 nm apart from each other. In this case the outside edge of the NW is to the left of the images; i.e. material thickness increases to the right. Figure 4.8b–d show the same structure after heating steps 7, 8 and 9 respectively from Table 4.1. Before heat is applied, the atomic columns for each of the $\{112\}$ twin facet can be clearly resolved. During heating the 3ML defect on the left is essentially stationary, while the 3ML defect on the right moves to meet it. In Fig. 4.8b, the columns become blurred/indistinct in the twinned region as the right side defect moves, and indicate the twin no longer occupies the full thickness of the NW along the electron beam direction. Although the blurring makes it difficult to observe the motion clearly, there is no indication of the defect splitting into its constituent $\frac{1}{6}\langle 112\rangle$ partials while moving. In the next frame (Fig. 4.8c), the twinned material section becomes even smaller until it finally disappears (Fig. 4.8d). In the final frame, all atomic columns are

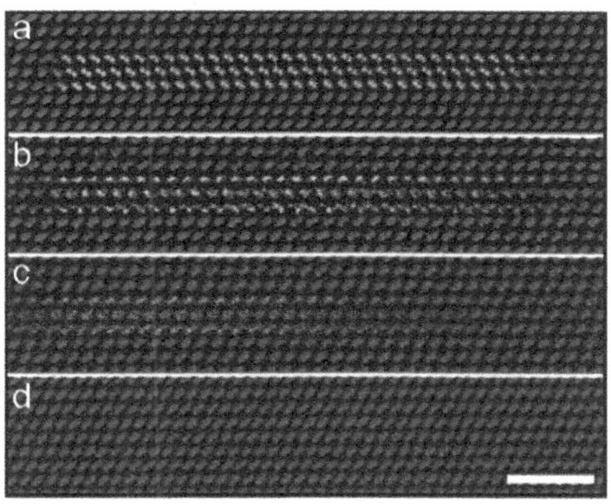

Fig. 4.8 Coloured band filtered ADF-STEM images of 3ML defects in a type 1 configuration (right) and type 2 configuration (left), during exposure to the increasing temperatures given in Table 4.1. **a** Before heating and after **b** step 7 (680 °C), **c** step 8 (680 °C), and **d** step 9 (700 °C), at which point the twin has been completely removed from the NW. Scale bar 2 nm. Adapted with permission from [8]. Copyright 2019 American Chemical Society

clearly resolved, again demonstrating the complete removal of the defects. This gives an example of a 3ML defect removal, and it also demonstrates the different forces acting on configuration 1 and 2 type defects. The left 3ML defect (configuration 2) shows very little movement, which contrasts the relatively rapid movement of the right defect (configuration 1). This matches with the forces from Fig. 4.3; in configuration 2 on the left forces F_{111} and F_{112} mostly cancel whilst on the right (configuration 1) the forces act together. A note to be made on the position of the defect, the left 3ML defect is approximately 32 nm from the edge of the NW. The position is obtained based on the intensity profile taken across the NW, as shown in Fig. 4.9. Profiles like this also allow for the determination of where the NW thickness becomes uniform where the image intensity flattens out.

The blurred appearance of defects in the process of moving observed in this heating experiment is consistent with simulated ADF-STEM images presented in Fig. 4.10. Figure 4.10a represents a model system produced of a GaAs NW with a 3ML which is kinked half-way along the defect line. Two models (not relaxed) were created, one with a single kink (Fig. 4.10b) and one with two adjacent kinks (Fig. 4.10e) at the midpoint along the defect line. Figure 4.10c, f shows ADF-STEM image simulations of these model structures, where a blurring similar to that seen in experimental images is observed, with examples presented in Fig. 4.10d, g. The extent

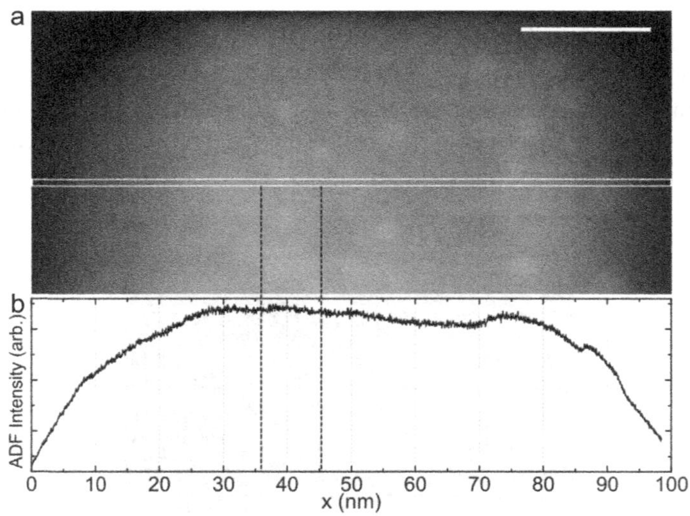

Fig. 4.9 **a** ADF-STEM image of NW 3 from Fig. 4.7 before any heating cycles. The dashed lines indicate approximately where the defect in Fig. 4.8 starts and ends. The white box indicates the region an intensity profile is taken from and is shown in (**b**). The position where the NW thickness becomes uniform due to the approximate hexagonal shape can be estimated from the point at which intensity flattens out, which in this example is around 30 and 75 nm from the left-hand side of the image. Scale bar 20 nm. Adapted with permission from [8]. Copyright 2019 American Chemical Society

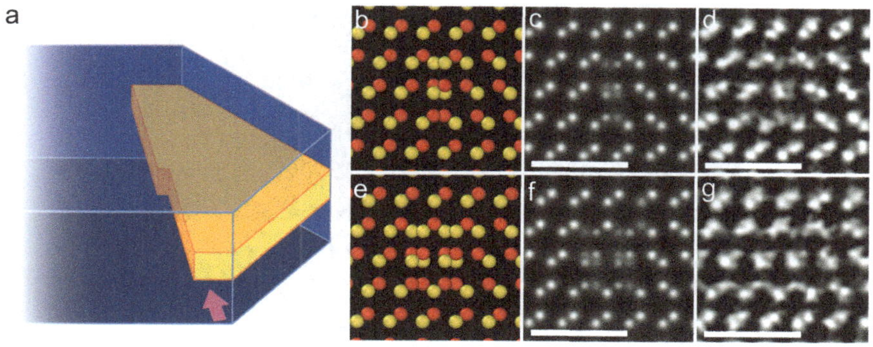

Fig. 4.10 **a** A 3ML defect in a NW with a kink half-way along the defect. **b** A model 3ML defect with a single kink in a GaAs NW viewed along the defect line direction, **c** an ADF-STEM simulated image of this view, and **d** an experimental image with similar appearance. **e** A model 3ML defect with two adjacent kinks in a GaAs NW viewed along the defect line direction, **f** an ADF-STEM simulated image of this view, and **g** an experimental image with similar appearance. Scale bars 1 nm. Model images generated using OVITO software [16]

of the blurred appearance will depend on the position of the kink along the defect line. The blurred appearance of defects in the process of moving is also consistent with similar observations made by Sarikov et al. [7] in molecular dynamics simulations of partial dislocation motion.

A similar experiment was performed using another NW after preparing a new heating chip with some more NWs from the same substrate. Instead of using programmed heat cycles and imaging in between cycles, the NW was imaged at a fast scan rate with temperature elevated to 570 °C. In this case, a 768×768 pixel image was recorded with a dwell time of 2 μs/pixel. A series of 618 images was acquired over a period of around 15 min. Images were aligned using SmartAlign software [17], a band filter applied and then a rolling average of 70 frames applied. Figure 4.11 shows three frames from this sequence of 618 images, and a full video is available in the supporting information of [8]. The first frame in 4.11a is an ADF-STEM image taken before any heat was applied. The image in Fig. 4.11b shows the first image from the stack of 618 images acquired live with temperature raised to 570 °C, and the image in Fig. 4.11c corresponds to the final image in the series. The initial frame from the image stack does not show atoms perfectly resolved since the defect started to move during the time it took to get the image in focus when live acquisition was started. The image series recorded showed a gradual progression from atomic columns being blurred, indicating the presence of a moving 3ML defect (Fig. 4.11b) to all atoms being perfectly resolved and of the same crystal orientation (Fig. 4.11c), indicating the 3ML defects were removed from the system. The motion and defect removal was much slower than the previous example since a lower temperature was used (570 °C vs 700 °C).

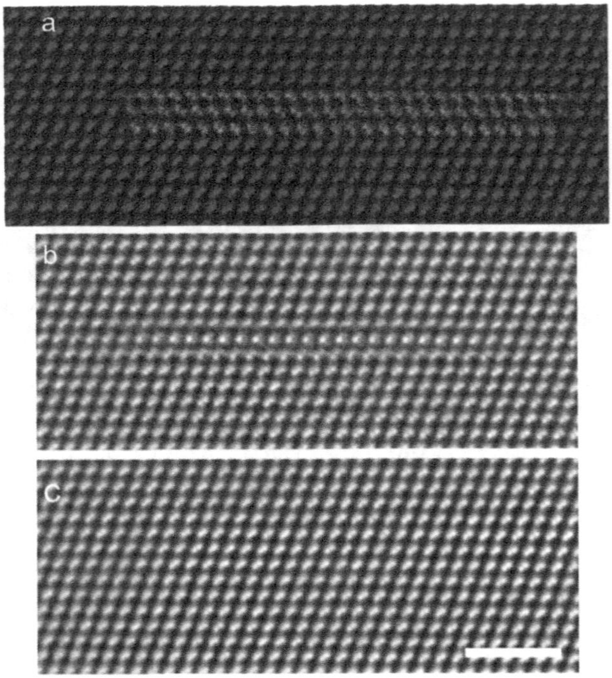

Fig. 4.11 **a** Coloured band filtered ADF-STEM image of an area with a 3ML high twinned region bounded by $\Sigma 3\{112\}$ twin facets at each end before any heating. **b** First frame from a live image series acquired during heating with the region appearing blurred as the defects move. **c** Final frame from the live image series which shows no blurred region and so indicates complete removal of the defects. Scale bar 2 nm. Adapted with permission from [8]. Copyright 2019 American Chemical Society

4.3.2 Movement Stopped by Interactions

A defect motion behaviour observed that differs to the defect removal seen in the previous section was defect motion that is eventually stopped by interaction with another defect, locking the new configuration in place. It has recently been reported using molecular dynamics simulations that inserting partial dislocations on adjacent $\{111\}$ planes in SiC can result in the partial dislocations moving towards each other and subsequently forming a stable 3ML defect configuration [18].

A demonstrative experimental example of defect motion being halted by interaction with another defect is presented in Fig. 4.12. The area before heat is applied is shown in Fig. 4.12a. At the top an example of a configuration 3 defect is seen, with a 3ML twin that extends across most of the image, and a second 3ML step increasing the twin to a height of 6ML on the right. Towards the bottom left of the image there is another pair of 3ML defects that together make a 6ML high twin. Since the 6ML was expected to be a more stable configuration, it was taken as a reference point for

Fig. 4.12 Coloured band filtered ADF-STEM images of 3ML defects close to the centre of a GaAsP NW during exposure to increasing temperatures given in Table 4.1. **a** Before heating and **b–g** show frames from heat cycle step numbers 3–8. The twinned section of material towards the top of the image moves to meet a step structure where it becomes stable. Scale bar 5 nm. Adapted with permission from [8]. Copyright 2019 American Chemical Society

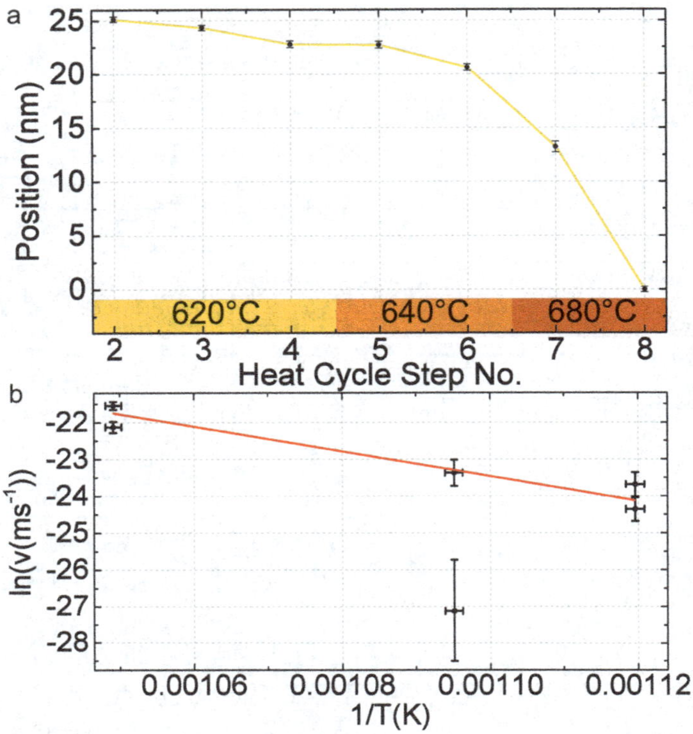

Fig. 4.13 **a** Position of a 3ML defect relative to final position during the heating experiment. **b** Measurement of the activation energy by fitting the data in part (**a**) to the linear form of Eq. 4.11. $Q = 2.9(6)$ eV. Adapted with permission from [8]. Copyright 2019 American Chemical Society

tracking the position changes in the other defect configurations. In the centre of the image is a 4ML twinned section of material that extends across the full width of the NW. The following frames in Fig. 4.12b–g shows the same area after heating steps 3–8 from Table 4.1. The lower pair of 3ML defects quickly form a stationary 6ML defect. The upper 3ML defect becomes blurred and moves to the right, reducing twin boundary area. The 3ML step at the top right barely moves (as may be expected, since it makes no change in twin boundary area by doing so, and is an example of a 3ML step in region (ii) from Fig. 4.1b. When the two 3ML defects meet and form a 6ML facet, the motion completely stops, which is consistent with a low mobility of higher {112} facets. The fact the columns are clearly resolved again indicates that a new stable configuration has been reached.

The position of the upper 3ML defect was tracked at each temperature cycle, and the position of the defect relative to final position is plotted in Fig. 4.13a for each of the frames shown in Fig. 4.12. Since heat cycles with a programmed time of elevated temperature were used, then values for velocities can be obtained for different temperatures. A linearised plot of velocities is shown in Fig. 4.13b, and a fit of the linear form of Eq. 4.11 gives a value of $v_0 = 4 \times 10^5$ m s^{-1} and $Q = 2.9(6)$ eV. 98 defects were tracked in total (including those close to the edges of the NW), and from extracting activation energy in this way the values of Q varied considerably from 1.8 to 7.6 eV. These unreliable values show how the forces acting on the 3ML defects vary strongly with their position, and a correct calculation requires detailed knowledge of the exact shape of the NW. In addition to uncertainties about NW shape there are several complicating factors that make quantitative study difficult. Factors include the different core structure of up- and down-steps; the variation of stacking fault energy with temperature [19]; pinning by impurity segregation [13, 15, 20]; varying composition in a core-shell structure; changes to the NW surface during heating [21]; and the effect of the electron beam [22–24]. In the absence of this information, it can only be said that these measurements give an upper bound to the activation energy of approximately 2 eV. These observations of 3ML defect motion are limited by time and temperature regimes that can be accessed inside the microscope. Since many defects only have a short distance available to travel before reaching the NW surface, only a few measurements are possible. Extraction of an activation energy from the data is also difficult due to the changing forces the 3ML defects experience as they move through the NW. In cases where there are individual defects close to the axis of the NW that travel a long distance, there is an observable increase in velocity with increasing temperature. If the defect is in region (ii) of the NW with constant thickness, then the prefactor v_0 in Eq. 4.11 should be constant.

4.3.3 No Movement: Stable Configurations

Although an example of a stable configuration has been presented in the previous Sect. 4.3.2, since roughly 25% of the defects did not move at all during the heating experiment, further examples ares shown in this section. The first demonstrative example is a 3ML step type located in region (ii), i.e. the thickness of the NW is constant, and an example that showed no observable movement after all heat cycles is shown in Fig. 4.14.

Another example of a stable defect, much higher than a typical 3ML defect, is presented in Fig. 4.15. It is a twinned Section 9ML high, with two additional steps along the top twin interface. The frames in Fig. 4.15b–d confirm that there is no change in the structure after heat cycles 8, 12 and 13 respectively. This represents another case of stable {112} facets with no movement of most of the facets involved in such configurations. Only the top step facet does begin to move at the higher temperatures and is removed after a long exposure to heat.

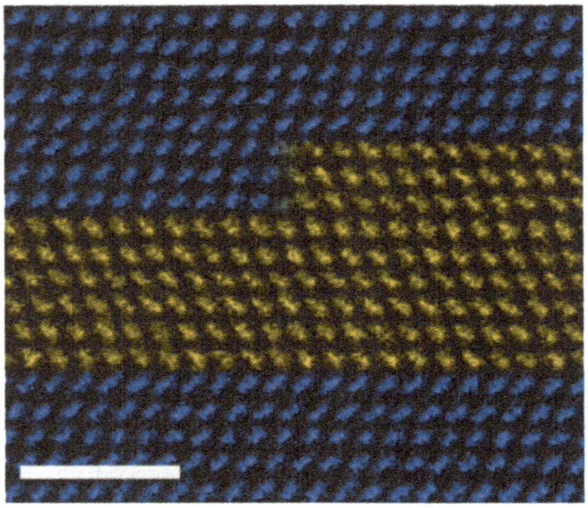

Fig. 4.14 Coloured band filtered ADF-STEM image of a 3ML step seen in the central area of a NW. After all stages of heating, the step does not move. Scale bar 2 nm. Adapted with permission from [8]. Copyright 2019 American Chemical Society

4.4 Velocity Analysis

To build a better understanding on the defect behaviours during heating cycles, similar analysis was carried out for 96 defects found in the three $GaAs_{1-x}P_x$ NWs shown in Fig. 4.7. The number of mobile defects for each heat cycle and their velocities were analysed. The behaviour of the 3ML defects is summarised by the charts in Fig. 4.16. The number of mobile (yellow) and stationary (blue) 3ML defects observed in the three NWs and the velocity distributions for the different temperatures are presented in Fig. 4.16a, b respectively. Each of these measurements are split into datasets for each NW in Fig. 4.16c, d to see how the trends vary between NWs. In the first temperature cycle at 620 °C, the vast majority (82%) of 3ML defects do not move. However, at 640 °C, 60% of the defects become mobile, and a significant number of defects begin to recombine or reach the NW surface after cycle 4. After two cycles at 640 °C, more than half of the initially mobile 3ML defects have been lost. While the number of immobile defects does not change after the second cycle at any temperature, an increase in temperature causes previously stationary defects to move. At the end of the experiment, 65% of 3ML defects have been removed; however almost 15% of the remaining defects remain mobile, indicating that 70% of defects could be removed by a slightly longer anneal.

At 620 °C the fastest-moving defect reaches a velocity of 0.68 nm s^{-1}, while at 640 °C some 3ML defects reach 0.76 nm s^{-1}. At higher temperatures still the maximum speed does not change significantly, but greater numbers of defects move at moderate speeds (0.2 to 0.3 nm s^{-1}). Looking at each NW individually, NW1 with a thinner and more tapered tip shows faster removal of the 3ML defects, with some of them trapped after initial movement. NW2 with the highest number of defects present of the three NWs, shows the most activity during the heating cycles. NW3 contains the most stable defect configurations, suggesting a relatively complex structure with many lock-type features present or many multi-height 3ML defects. There is not much

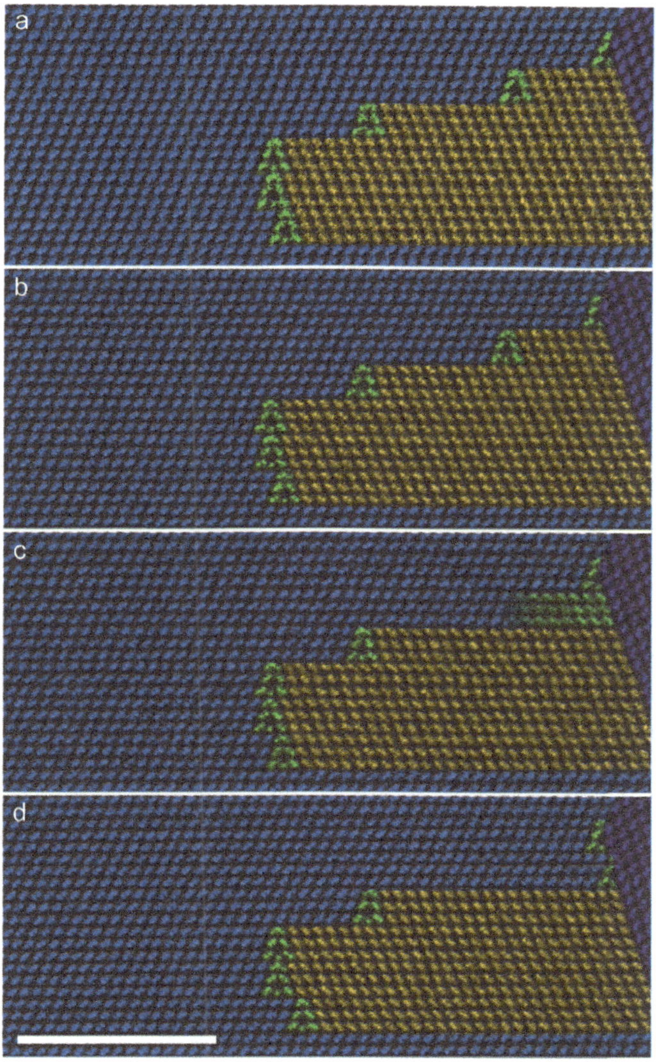

Fig. 4.15 Coloured band filtered ADF-STEM images showing a type 4 configuration where even at high temperatures most of the {112} twin facets do not move, only the top one does. Images are taken after heating cycles in Table 4.1 **a** before heating **b** cycle 8 **c** cycle 12 **d** cycle 13. Scale bar 5 nm. Adapted with permission from [8]. Copyright 2019 American Chemical Society

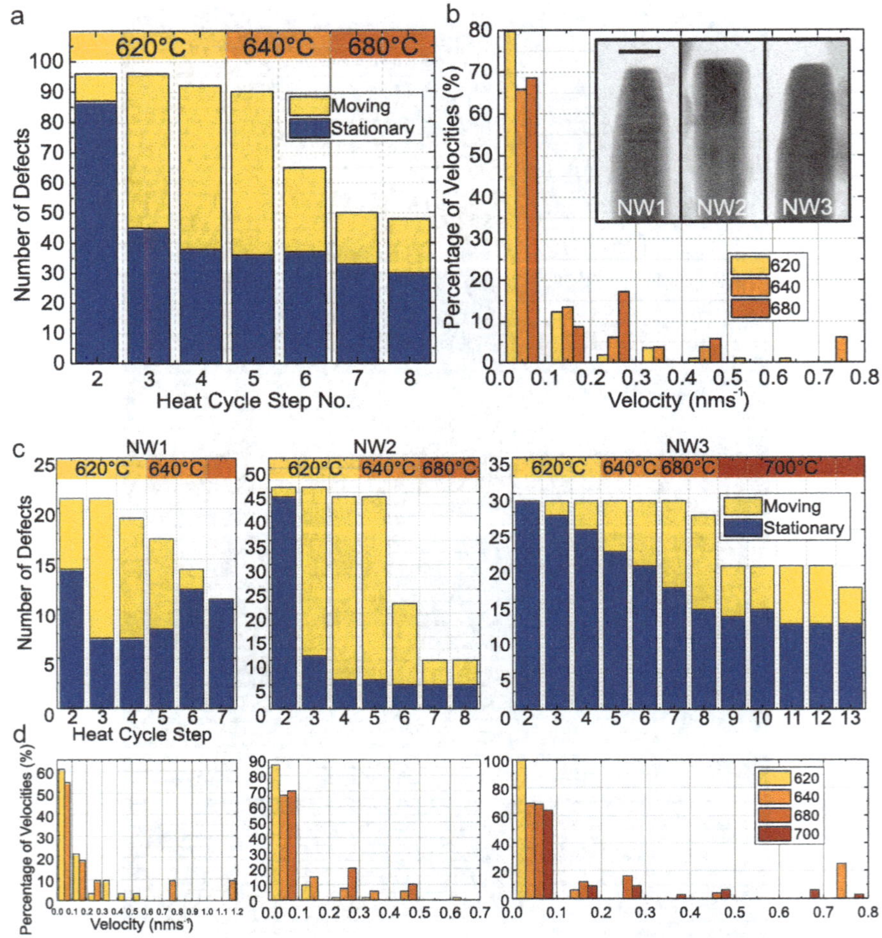

Fig. 4.16 **a** Number of defects observed (blue = stationary, yellow = mobile). **b** Histograms of defect velocities normalised to total number of velocities measured for each temperature observed during the temperature cycles given up to cycle number 8 in Table 4.1. The inset shows BF-STEM images of the 3 NWs used, scale bar 100 nm. **c** The number of defects observed separated to each NW at different heat cycle steps. **d** Histograms of defect velocities for each NW normalised to total number of velocity measurements made for each temperature cycle step. Adapted with permission from [8]. Copyright 2019 American Chemical Society

difference in the defect velocities between the three $GaAs_{1-x}P_x$ NWs observed. Higher temperatures lead to greater velocities, as expected. The lack of higher defect velocities is because the fast-moving defects either are removed when they reach the NW surface or stop moving when they form a stable configuration in the first few temperature cycles. The wide range of velocities indicates the variety of forces that 3ML defects experience, which is consistent with the calculations from Sect. 4.2. The relatively low velocities (dislocations in bulk GaAs experiencing similar forces reach velocities of several thousand nm s^{-1} [13]) are consistent with the kink nucleation model and the low velocities observed for defects in thin films [25].

4.5 Motion of More Complicated Defect Configurations

This section looks at some additional examples of motion behaviour that is more complicated than those previously demonstrated. Figure 4.17a presents an area containing 11 labelled defects after heat cycle 2 in Table 4.1. Figure 4.17b–d correspond to the same area after heat cycles 7, 8, and 11. The following list describes defects and their motion observed in Fig. 4.17.

- Defects 1 and 2—An up/down pair of 3ML defects. These defects act in a similar way to the example seen in Fig. 4.8, where once motion has started, the 3ML on the right side moves faster than the 3ML on the left, with both eventually removed from the system. One difference is the motion required higher temperature to initiate in comparison with the example in Sect. 4.3.1. In Fig. 4.17 the motion was initially delayed by defect 2 and defect 3 locking each other in place. Once motion was initiated, movement is very fast.
- Defect 3—A 3ML defect on a $\left(11\bar{1}\right)$ plane. When defect 2 starts to move, defect 3 changes position at a slower speed. Defect 2 is likely to be a configuration 1 type defect with larger net forces acting on it, while the forces acting on defect 3 are mostly of a configuration 3 type. There is a component of a F_{112} force acting on defect 3 along the direction of the $(11\bar{1})$ plane it glides on. This component will be smaller because the $(11\bar{1})$ plane direction is at an approximately 70° angle from the horizontal direction F_{112} points, since F_{112} always points towards the edge of the NW (as was seen in Sect. 4.2). Once defect 3 moved down far enough from defect 2, a portion of the $\left(11\bar{1}\right)$ twinned structure (purple phase) is transformed into the same crystal structure as the surrounding material (blue phase), with a 5ML high section changing orientation. The motion of this defect creates a new 3ML defect at the bottom of the image. This is a rare example recorded where an inclined 3ML defect moves, and demonstrates behaviour that is similar to a configuration 3 defect.
- Defects 4, 5 and 6—Defect 4 is a 1ML defect interacting with the $\left(11\bar{1}\right)$ twin, and defects 5 and 6 are 2ML defects interacting with the $\left(11\bar{1}\right)$ twin. Once defect 3 has moved, defects 4, 5 and 6 initiate their changes. Defect 4 is very quickly removed from the system, while defects 5 and 6 become a new 3ML and 2ML defect that

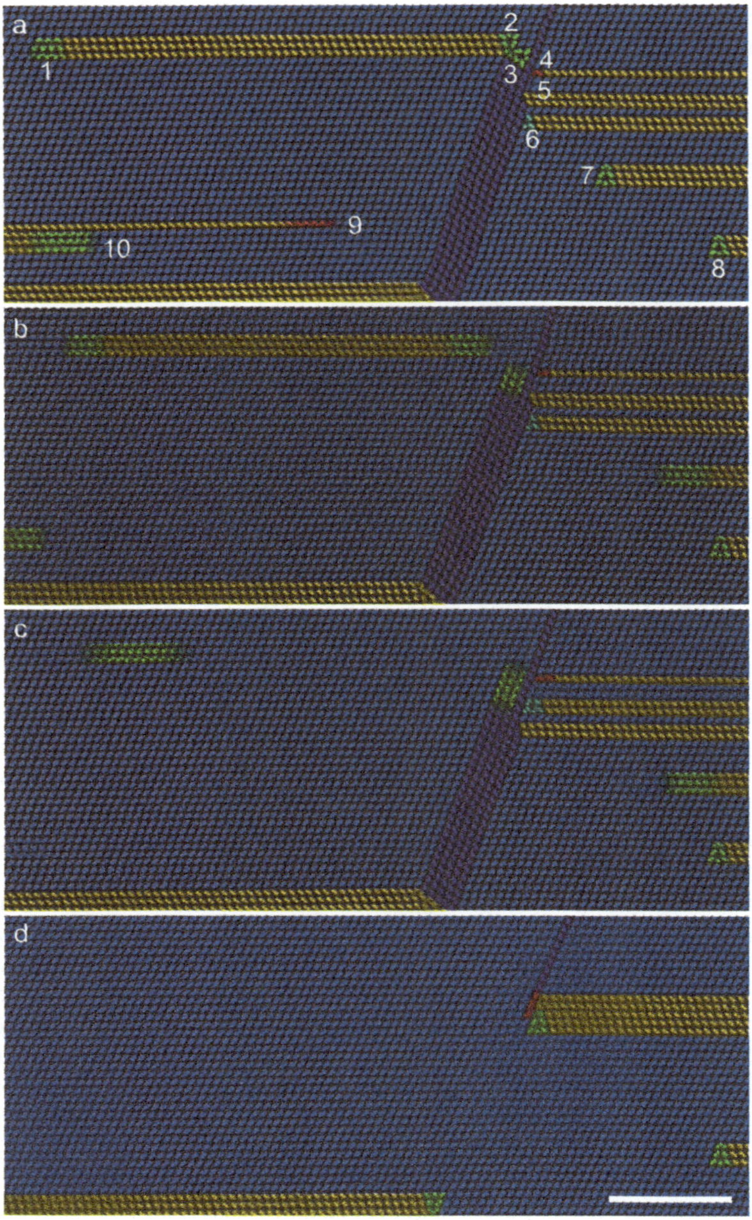

Fig. 4.17 A series of coloured band filtered ADF-STEM images taken from NW3 that show progression of defect motion for a variety of defects numbered in (**a**). **a**–**d** show images after heat cycles 2, 7, 8, and 11 respectively. Scale bar 5 nm

interact with the one remaining $(11\bar{1})$ twinned region. This demonstrates how defect interactions have a strong influence on their mobility, with rapid motion occurring once the defects have been provided sufficient energy to overcome the activation energy barrier to initiate motion.

- Defect 7 and 8—Two 3ML defects. Defect 7, initially atomically resolved shows a gradual movement typical for a configuration 3 defect. Defect 8 remains stationary throughout, suggesting an interaction with another defect not visible here.
- Defect 9—A 1ML defect in the process of moving. This defect is very quickly removed from the system, as expected for a non-zero Burgers vector defect.
- Defect 10—A 3ML defect in the process of moving. This defect gradually moves out of frame, and moves faster at higher temperatures.

Another example of complicated defect interaction is presented in Fig. 4.18. The frame in Fig. 4.18a was recorded in an area of NW1 before heating. Defects 1 and 2 are a 5ML high step formed by the combination of a 3ML and a 2ML defect.

Defect 3 is a 2ML step facet, and defect 4 is a 2ML section of twinned material. As the heat cycles progress, defect 4 begins to move to the left, in a way that can be described as a configuration 3 type defect. The motion of defect 4 also pulls defect 3 along with it. Another 2ML defect appears (defect 9) from the left side of the frame in Fig. 4.18b, which progresses towards defect 4 (Fig. 4.18c) until they meet and are both removed from the NW. After defects 4 and 9 have been removed, a very rapid change of structure occurs, and involves what is assumed to be defects that cannot be seen in the earlier frames now moving into the frame.

In Fig. 4.18d, it is seen that defects 1 and 2 have become two 3ML defects, a dipole of 1ML defects, and a further 3ML defect. Defect 3 has become the base of the lower 3ML defect in defect 10. It is assumed a 2ML defect from out of frame on the left side has turned defect 2 into a new 3ML defect plus a 1ML defect. Similarly, it is assumed a 2ML defect which terminates the 2ML twinned section above defect 3 has moved from the left and has combined with defect 3. The dramatic change in structure has resulted in a reduction of Burgers vectors. All of the defects contributing to defect 10 have a net Burgers vector of zero. This new structure also appears to be very stable, since after 4 more heat cycles at higher temperatures, the only change seen in Fig. 4.18e is the defects labelled 11 shift by one atom towards the right. The stability comes from a mixture of being a step type defect and from being a relatively tall defect.

The motion of defects in Fig. 4.18 is faster than the motion of 3ML defects observed, as expected for defects with a non-zero Burgers vector and a shorter height. These defects with a non-zero Burgers vector will experience the configuration 3 F_{111} forces in addition to surface image forces as described in Sect. 1.3.

Defect 5 is an example of a configuration 3 type 3ML defect, and moves in a similar way to previously seen configuration 3 type defect motion. Defect 6 is a 3ML step interface which shows no sign of motion. Defect 7 is a 3ML defect interacting with a Hirth lock and does not move. Defect 8 is a LCL that does not move.

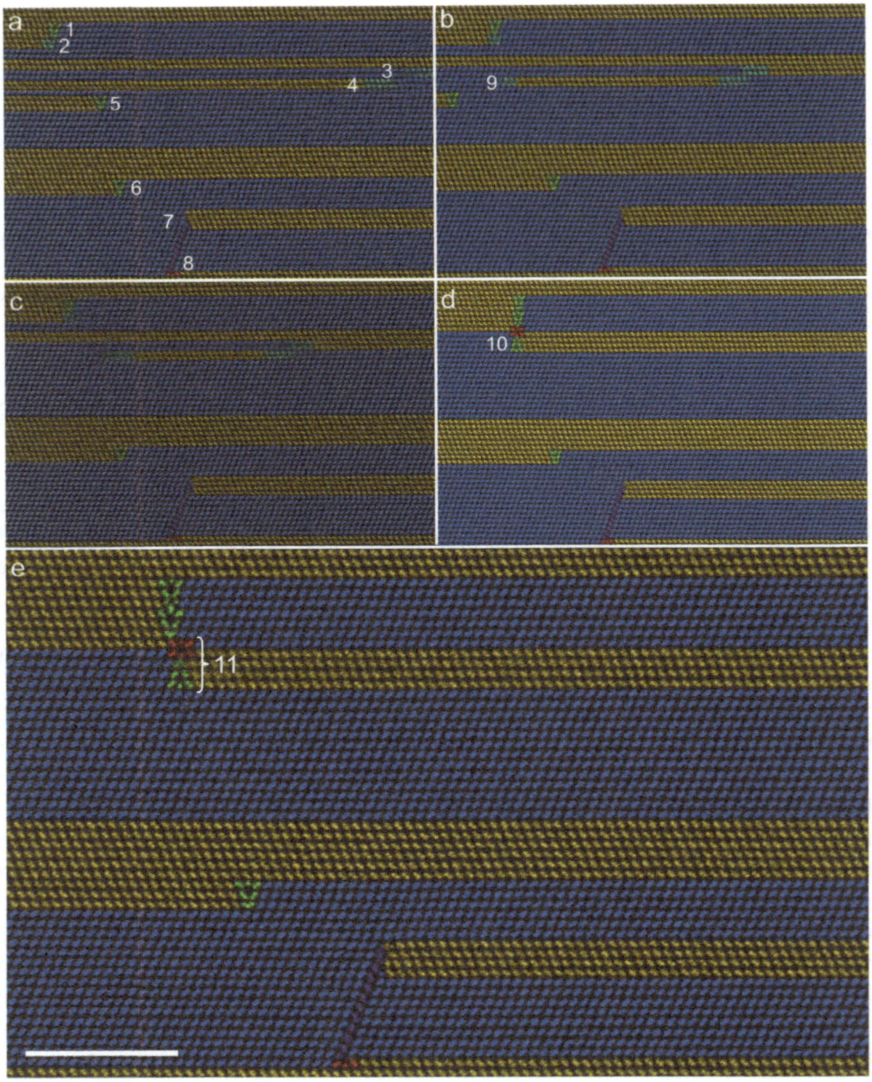

Fig. 4.18 A series of coloured band filtered ADF-STEM images taken from NW1 that shows progression of defect motion for a variety of defects numbered in (**a**) and (**b**). **a** Image from before heat is applied and **b–e** show images after heat cycles 2, 3, 4, and 8 respectively. Scale bar 5 nm

The transformation observed in Fig. 4.18 demonstrates that drastic structural changes can occur in a NW by annealing a defective structure, and the observed changes tend to improve material quality.

4.6 Chapter Summary

This chapter has explored how some of the defect types that were seen in Chap. 3 move when the system is annealed. Most attention has been given to the $\Sigma3\{112\}$ (3ML) defect since as a zero Burgers vector defect, it would not normally be expected to move. It was seen how there are a range of different possible configurations for these 3ML defects to exist as inside a NW structure, and the exact configuration of the defect will dictate what forces act on the defect and drive them into motion. The forces acting on 3ML defects come from a reduction in the size of twinned material area which reduces the Gibbs free energy of the system.

By using a series of programmed heating cycles, a systematic study was performed on some NWs where 96 individual defects were tracked. Defects were seen to either 1—be completely removed from the system, 2—begin to move and subsequently stop once they get locked in place or 3—show no motion since they are already in a stable configuration. Some examples were also seen of motion of defects with a non-zero Burgers vector, and showed expected behaviour that is different to 3ML defects. The majority of 3ML defects were observed to be unstable in a NW when exposed to high temperatures. Movement of 3ML defects is thermally activated, and they can be removed by post growth annealing at temperatures above $\approx$640 °C. By using different temperatures, velocities of 3ML defects were measured and an estimate for the upper bound of activation energy was obtained and found to be $\approx$2 eV. The velocities found are consistent with those expected from kink-nucleated glide of dislocations.

While roughly 25% of the defects observed did not move at all and were truly stable, it was seen that 70% of defects present in the studied system could be removed, and suggests that annealing after growth is a relatively simple method to improve material quality in self-catalysed NWs. A strategy to improve material quality is to employ an anneal immediately after droplet consumption and before any shell growth. For the GaAsP system, the use of either long anneals at $\approx$570 °C for between 15 to 30 min or shorter anneals at $\approx$640 °C for between 5 to 10min are recommended to remove defects from NWs. The choice of anneal time will depend on NW radius, with shorter anneal times used for thinner NWs.

References

1. Heon Kim Y, Woo Park D, Jun Lee S (2012) Gallium-droplet behaviors of self-catalyzed GaAs nanowires: a transmission electron microscopy study. Appl Phys Lett 100:033117
2. Dastjerdi MHT, Boulanger JP, Kuyanov P, Aagesen M, LaPierre RR (2016) Methods of Ga droplet consumption for improved GaAs nanowire solar cell efficiency. Nanotechnology 27:475403
3. Sanchez AM et al (2017) Nonradiative step facets in semiconductor nanowires. Nano Lett 17:2454–2459
4. Zamani M et al (2018) Optimizing the yield of A-polar GaAs nanowires to achieve defect-free zinc blende structure and enhanced optical functionality. Nanoscale 10:17080–17091
5. Xu L et al (2008) Structure and migration of (112) step on (111) twin boundaries in nanocrystalline copper. J Appl Phys 104:113717
6. Sarikov A et al (2019) Molecular dynamics simulations of extended defects and their evolution in 3C–SiC by different potentials. Model Simul Mater Sci Eng 28:015002
7. Sarikov A, Marzegalli A, Barbisan L, Montalenti F, Miglio L (2019) Structure and stability of partial dislocation complexes in 3C-SiC by molecular dynamics simulations. Materials 12:3027
8. Gott JA et al (2019) Defect dynamics in self-catalyzed III-V semiconductor nanowires. Nano Lett 19:4574–4580
9. Glas F, Harmand J-C, Patriarche G (2007) Why does wurtzite form in nanowires of III-V zinc blende semiconductors? Phys Rev Lett 99:146101
10. Gottschalk H, Patzer G, Alexander H (1978) Stacking fault energy and ionicity of cubic III-V compounds. Physica Status Solidi (a) 45:207–217
11. Glas F, Ramdani MR, Patriarche G, Harmand J-C (2013) Predictive modeling of self-catalyzed III-V nanowire growth. Phys Rev B 88:195304
12. Hirth J, Lothe J (1982) Theory of dislocations. Krieger Publishing Company
13. Sumino K (1994) Mechanical behaviour of semiconductors, in Handbook on semi-conductors, pp 73–182. Elsevier Science, Amsterdam
14. Jones R, Oberg S, Marklund S (1981) Structure and energy level calculations of dislocations in gallium arsenide. Philos Mag B 43:839–852
15. Yonenaga I, Sumino K (1989) Impurity effects on the generation, velocity, and immobilization of dislocations in GaAs. J Appl Phys 65:85–92
16. Stukowski A (2010) Visualization and analysis of atomistic simulation data with OVITO-the open visualization tool. Model Simul Mater Sci Eng 18
17. Jones L et al (2015) Smart Align—a new tool for robust non-rigid registration of scanning microscope data. Adv Struct Chem Imaging 1:8
18. Scalise E et al (2020) The origin and nature of killer defects in 3C–SiC for power electronic applications by a multiscale atomistic approach. J Mater Chem C 8:8380–8392
19. Moon W-J, Umeda T, Saka H (2003) Temperature dependence of the stacking-fault energy in GaAs. Philos Mag Lett 83:233–247
20. Yonenaga I, Sumino K (1987) Effects of In impurity on the dynamic behavior of dislocations in GaAs. J Appl Phys 62:1212–1219
21. Chou Y-C, Panciera F, Reuter MC, Stach EA, Ross FM (2016) Nanowire growth kinetics in aberration corrected environmental transmission electron microscopy. Chem Commun 52:5686–5689
22. Chin AK, Keramidas VG, Johnston WD, Mahajan S, Roccasecca DD (1980) Evaluation of defects and degradation in GaAs-GaAlAs wafers using transmission cathodoluminescence. J Appl Phys 51:978–983
23. Maeda K, Yamashita Y, Maeda N, Takeuchi S (1990) Radiation enhanced dislocation glide and rapid degradation. MRS Proc 184:69
24. Lavagne S, Levade C, Vanderschaeve G (2006) Transmission electron microscopy in situ investigation of dislocation behaviour in semiconductors and the influence of electronic excitation. Philos Mag 86:4923–4940
25. Maeda K, Yamashita Y (1993) Dislocation motion in strained thin films. Are Kinks colliding with each other? Physica Status Solidi (a) 138:523–532

Chapter 5
Interfaces in Nanowire Axial Heterostructures

5.1 Introduction

Axial heterostructures have gained a great deal of attention in NWs as they are crucial building blocks for semiconducting optoelectronic devices. The unique control of the VLS growth can be used to create confined quantum systems inside NWs, and single QDs to multiple QD systems are being used for photon sources [1] and in laser applications [2]. The ability to control physical dimensions, position and composition of these quantum structures inside NWs means that optical properties can be engineered and precisely tuned. In order to obtain a spectrally sharp emission from QD structures, sharp interfaces between QD and the barrier material are required. The ideal configuration would be a perfectly symmetrical composition profile shape with a step function at the interfaces. Interfaces in axial heterostructures in NW systems are rarely seen to be atomically sharp, and there is usually a gradual change in composition [3]. This is primarily a result of the so called '*reservoir effect*', i.e. the residual element concentration left in the catalyst droplet once the source flux is switched off or changed [4].

The VLS growth mechanism is different to the traditional VS growth in thin films. Some attempts to model growth of axial heterostructures in NWs can be found in the literature [4–7]. This section explores the suitability of these theoretical models to explain experimental data obtained by electron microscopy. While these models have been applied to other systems previously, this study looks at the GaAsP/GaAs/GaAsP system in NWs, for which the interface sharpness has not been analysed previously. A sample with variable NW radius was grown to explore interface sharpness dependence with NW radius. Additionally, the change of QD size with radius of NW was also analysed.

J. A. Gott, *Defects in Self-Catalysed III-V Nanowires*, Springer Theses,
https://doi.org/10.1007/978-3-030-94062-1_5

5.2 GaAsP-GaAs-GaAsP Quantum Dots in Self-Catalysed Nanowires

The results in this chapter are largely based on just one growth, and a schematic of the growth design is shown in Fig. 5.1a. Different NW radii were achieved by following the growth procedure described by Zhang et al. [8]. The structure is a GaAsP NW with repeated insertions of 5 GaAs QDs of increasing size. This heterostructure was achieved by using precisely controlled changes in material fluxes for the times labelled in Fig. 5.1a. Figure 5.1b shows an ADF-STEM image of a NW with the five repeated groups of five QDs, with Fig. 5.1c showing a magnified view of the QDs labelled as group 1 in Fig. 5.1b. Figure 5.1d, e present a SE SEM image and an in-lens SEM image of the same NW growth still attached to substrate. The SEM images show an example of NWs forming bundles after exposure to a scanning electron beam, where the NWs stick together by van der Waals forces [9]. NW diameter varies approximately in the range 20–75 nm.

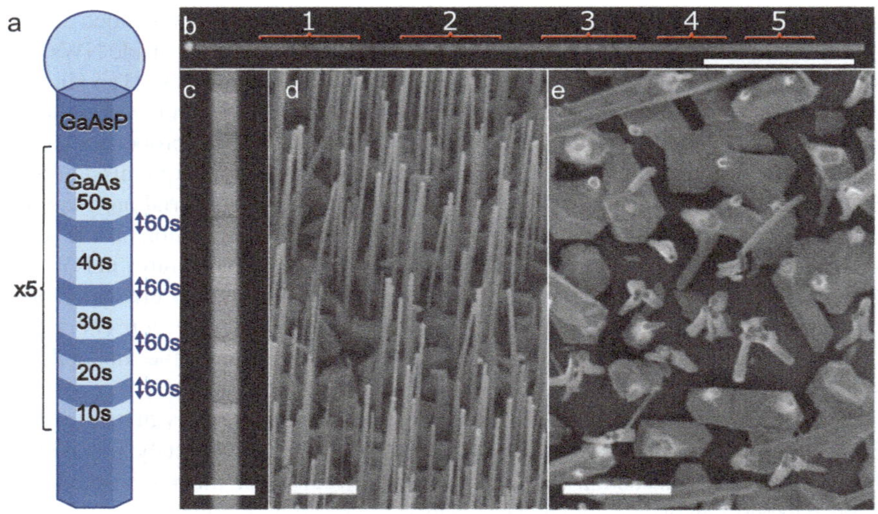

Fig. 5.1 **a** Schematic of a sample grown with NWs of varying diameters with QDs inserted axially along the NW. **b** A low magnification ADF-STEM image of a NW from the grown sample and shows the sample has the designed pattern of five sets of five QDs. **c** Higher magnification ADF-STEM image of one set of QDs. **d** SE SEM image of the sample on the substrate (accelerating voltage 5 kV). **e** A top-down in-lens SEM image of the sample on the substrate (accelerating voltage 1 kV). Scale bar in **c** 100 nm. Scale bar in **b**, **d** and **e** 1 µm

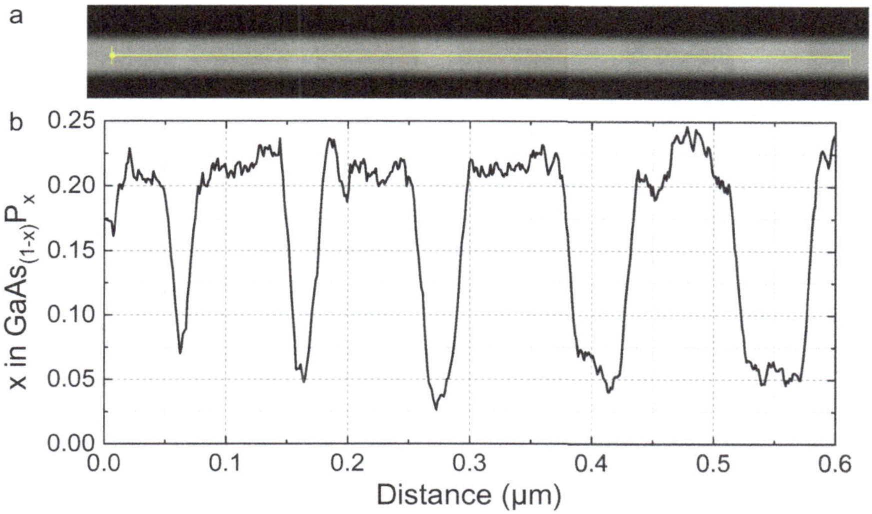

Fig. 5.2 Example EDX data used to calibrate composition profiles. **a** ADF-STEM image of a group of QDs. The yellow line indicates where an EDX linescan was taken. **b** P composition profile from the EDX linescan. Quantification is based on the Cliff-Lorimer ratio technique using calculated k-factors

5.2.1 Converting ADF Intensity to Composition

To extract quantitative information from ADF images, a method to correlate the intensity of ADF-STEM images with the semiconductor composition is required. The terminology in this section refers to ADF images, and the experimental settings of the microscope (detector and camera length) means electrons scattered to angles in the range (45 ± 5) to (180 ± 8) mrad are detected. This scattering angle range corresponds to the commonly referred to high angle scattering region (50–200mrad) that form HAADF images [10–12]. Since this section deals exclusively with $GaAs_{(1-x)}P_x$ ternary compounds the ADF intensity conversion method will be outlined for this system.

The method used in this section largely follows the procedure described by Priante et al. [5, 13, 14], where it is described how ADF intensity can be converted into a quantitative composition by using EDX measurements as a chemical calibration. Contrast between ADF intensities between two regions of compositions x_0 and x can be described by the ratio [14]

$$R = \frac{I}{I_0} = \frac{xZ_P^\alpha + (1-x)Z_{Ga}^\alpha + Z_{As}^\alpha}{x_0 Z_P^\alpha + (1-x_0)Z_{Ga}^\alpha + Z_{As}^\alpha} \tag{5.1}$$

where I and I_0 are ADF intensities from the two regions, Z_i is the atomic number of species i, and $\alpha \leq 2$ is a fitting parameter that accounts for deviations from

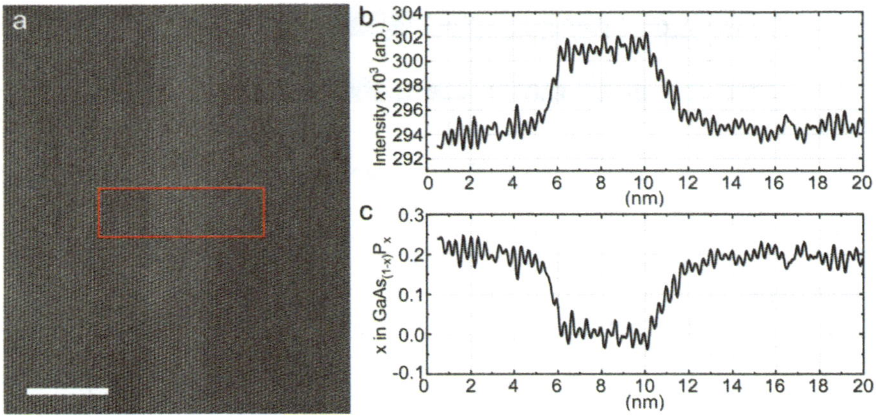

Fig. 5.3 Example process used to convert ADF-STEM intensity to P composition profile. **a** ADF-STEM image of a QD. The red box indicates the area where intensity profile is taken. Scale bar 10 nm. **b** Image intensity profile. **c** P composition profile produced from the intensity profile after processing the data

pure Rutherford scattering. To obtain an expression that correlates composition x to intensity I, a value for α is required. To get this value, two points are chosen from the image that is to be calibrated, and the intensity ratio R is calculated. Composition values x and x_0 are then taken from an EDX line profile of the same region, and an example line profile is shown in Fig. 5.2. The EDX spectrum is quantified using the Cliff-Lorimer ratio technique (see Sect. 2.5.2). A computational equation solver is then used to solve Eq. 5.1 for a value of α. For the profiles presented in this chapter, EDX values are taken to be $x_0 = 0 \pm 0.04$, $x = 0.2 \pm 0.04$, with these values representative of most of the NWs examined. Once an α value is found, Eq. 5.1 is re-arranged to

$$x = \frac{DR - Z_{As}^{\alpha} - Z_{Ga}^{\alpha}}{Z_P^{\alpha} - Z_{As}^{\alpha}} \tag{5.2}$$

$$D = x_0 Z_P^{\alpha} + (1 - x_0) Z_{Ga}^{\alpha} + Z_{As}^{\alpha} \tag{5.3}$$

An example of an intensity profile taken from an image, and converted profile is shown in Fig. 5.3. For the example shown here, points are taken from the profile in Fig. 5.3b, which shows a 12 point adjacent average line profile of the red box indicated in Fig. 5.3a. The ratio of intensity is 0.9804 and by solving Eq. 5.1 for α using $x_0 = 0 \pm 0.02$, $x = 0.2 \pm 0.02$, $\alpha = 0.2741$. While this process is not perfect, it is a simple and relatively easy method of obtaining composition profiles of NW axial heterostructures. An alternative method to quantify ADF intensity could be to use simulations (e.g. as in [15]), however this is not used in this thesis.

Nevertheless there are some factors that will influence this type of analysis, such as the background intensity on the image, with a background removal required for

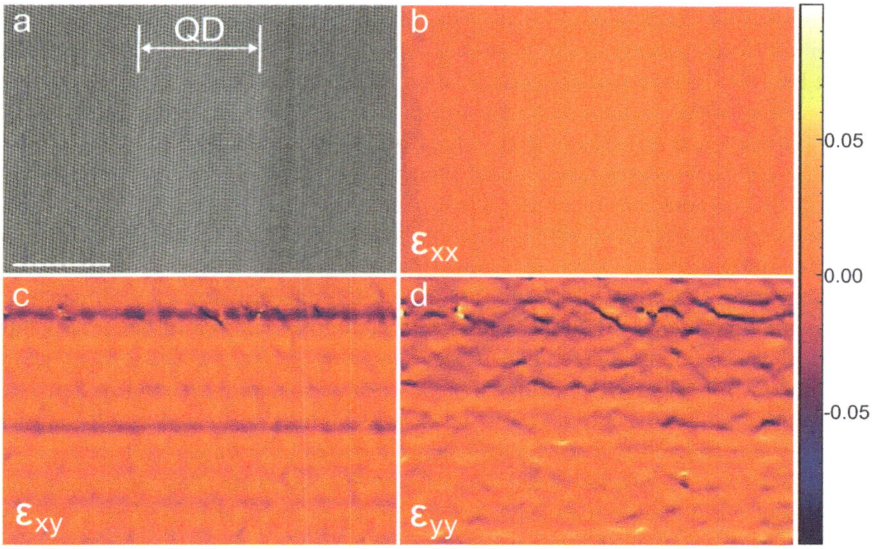

Fig. 5.4 Determining strain in a QD using GPA analysis. **a** ADF-STEM image of a QD, indicated by the labelled arrow and vertical lines. Scale bar 10 nm. **b–d** Strain components ϵ_{xx}, ϵ_{xy} and ϵ_{yy} respectively. Some strain is present in the QD structure

interface analysis. Strain in the system strongly influences the STEM intensity [12, 16], as explored in more detail in the introduction Sect. 1.4.5. An example high magnification ADF-STEM image of a QD and the accompanying strain map acquired from GPA is shown in Fig. 5.4. The ϵ_{xx} component indicates a small amount (<2%) of strain present in this system. The horizontal features present in the strain maps are attributed to image distortion caused by sample drift.

As was seen in Sect. 1.4.5, While strain is known to influences contrast in ADF-STEM imaging [16], surface relaxation can also influence contrast [12]. Relaxation of the structure is accompanied by the bending of lattice planes, reducing intensity at high scattering angles due to dechannelling of electrons from atom column positions. In ADF imaging of QD structures, this effect can lead to a dip in the intensity profile [12].

If strain is present in the sample, like in Fig. 5.4, then there is another very important parameter to consider in the quantification analysis based on intensity profiles. Sample alignment with respect to the electron beam direction is extremely important, i.e. sample tilt is a crucial factor. For a heterostructure like a QD, with lattice strain due to the mismatch with respect to the barrier material, any tilt of the atom columns away from the electron beam direction results in a shift in image intensity [17]. ADF intensity in zone axis conditions is strongly influenced by channelling. The plane bending due to strain relaxation at the sample surface will cause dechannelling, and results in a decrease in intensity at either side of the QD. If the structure is tilted slightly off zone-axis conditions, i.e. atom columns are away from the electron beam direction, the dechannelling effect is either compensated for or enhanced on each side of the heterostructure. The appearance of an interface intensity profile will depend on tilt angle of the atom column with respect to the electron beam, generating a dip in intensity with a height and distance from interface that depends on tilt angle. A higher tilt angle results in a greater change in intensity of the dip compared to zone-axis conditions.

In heterostructures, with different materials at each side of the interface the change in composition is almost never a step function, with compositional segregation. This will also affect the intensity line profile of any interfaces. Due to the influence of segregation, strain and tilt angle, a reliable quantitative analysis of heterostructure interfaces based on just image intensity becomes very difficult. An example demonstrating how important these factors are in the NW system is shown in Fig. 5.5. This figure corresponds to a group of five QDs in a single NW imaged at different tilt angles. The different QDs are indicated by dashed coloured lines in the figure. To show the different tilts used in the images, CBED images are shown to the left of the NWs in Fig. 5.5a, where the central QD is used to obtain the CBED image. The NW labelled (2) corresponds to an on-axis image with the central QD used to align the sample. A schematic of sample orientation is shown in Fig. 5.5b and line profiles along the five QDs are shown in Fig. 5.5c. This image is taken as a reference to analyse the influence of the tilt on the QD interfaces. The central QD in the on-axis orientation (2) shows a slight dip in intensity at each interface on the intensity line profile. This is similar to the dips in intensity previously observed in thin film samples and explained by strain in those systems [12, 17]. Small tilts, such as orientations (1) and (3), produce dramatic changes to the line profile for the central QD. For tilt (1) the intensity increases on the left while decreasing on the right, enhancing the strain dip. The opposite trend is seen when the NW is tilted the other way (3). Note that while the central QD profile would indicate that the NW is in zone (label (2) in Fig. 5.5), the rest of the QDs on either side show variation in intensity similar to the change seen for different tilt angles, generating dark/bright bands at the QD interfaces in the ADF images. This suggests the NW is bending. When the NW is tilted, a similar change in line profile as was seen for the central QD is seen with the other QDs, with the interface with a higher intensity changing from one side to the other when tilted.

This demonstrates how important sample tilt is for quantitative analysis of interface intensity profiles, and this can be further complicated if the NWs are bent. This also demonstrates that strain and tilt effects on intensity previously reported in thin films also apply to NWs.

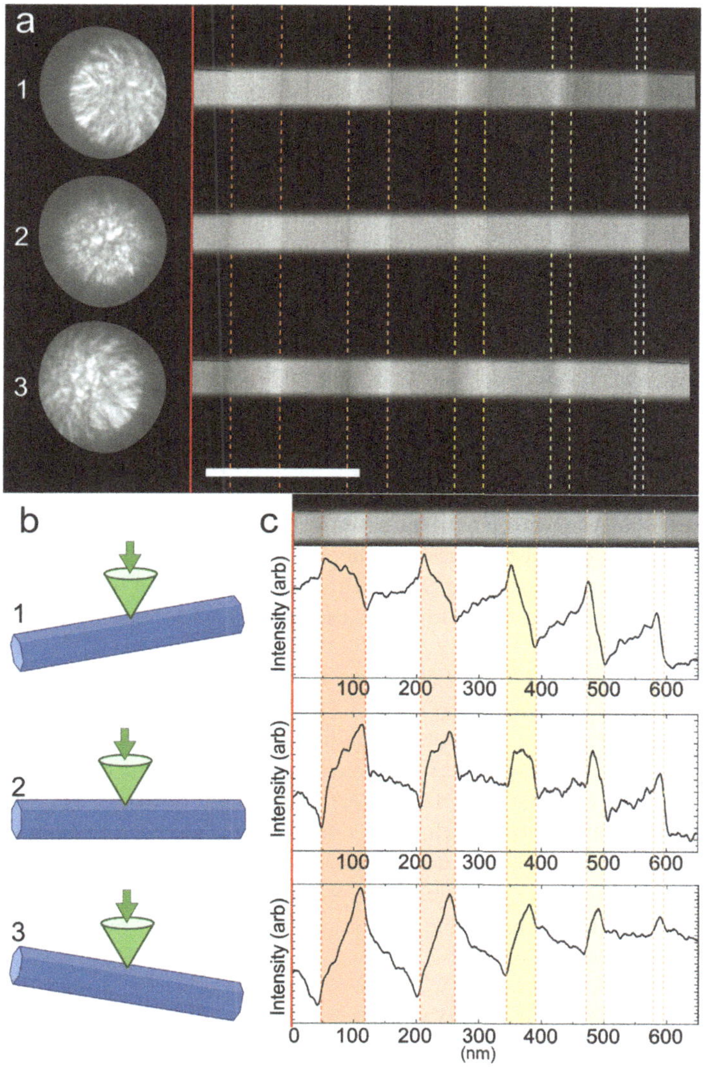

Fig. 5.5 **a** A series of CBED images numbered 1–3 for different sample tilts shown schematically in **b** along with beam direction. CBED image 2 indicates the sample is on axis. Next to each CBED image is the corresponding ADF-STEM image of the same group of QDs. The dashed lines act as a guide indicating QD location. Scale bar 200 nm. **c** Plots of image intensity taken from the images in **a**. Coloured bands and dotted lines act as a guide for position of QDs along the NW

5.3 Interface Models

This section looks into some theoretical models to describe the interfaces of axial heterostructures in NWs. Four models are considered, two of these are models recently reported specifically for NWs, one reported by Dubrovskii [4] and one by Priante et al. [5]. The Muraki model has also been considered, since it has widely been used to analyse interface segregation in SCs and finally an empirical model is considered. The models looked at here were explored in more detail in the introduction (Sect. 1.4). The model developed by Dubrovskii is considered first since this model considers the details of NW growth in the most detail. The models are then all considered together and an assessment is made on how well each model fits experimental data.

5.3.1 Dubrovskii's Kinetic Model

This model is a relatively complicated model with many parameters. Not all of the parameters have known values and there is no analytical relation $x(z)$ between composition x and position along NW z. This means this model is not ideal for fitting data and finding parameters, however for the purpose of comparing models, a fit is made using this model.

In order to get a relation between composition x and NW height z, full profiles of $a_{\mathrm{P}}(t)$, $a_{\mathrm{As}}(t)$ (Eq. 1.22) and $z(t)$ (Eq. 1.20) are required for different sections of growth (see Table 1.3 for parameter definitions). To fit data using the model from Sect. 1.4.1, the growth was split into three time sections (in a similar way to Fig. 1.14 in Sect. 1.4.1), 1—before QD, 2—QD growth and 3—after QD growth which is representative of sample growth. The composition profile is then built using Eqs. 1.22, 1.20 and 1.21. A model fit was produced by writing a profile builder in python and using the scipy curve_fit routine to optimise parameters. Some starting values were chosen (QD1 column, Table 5.1) based on values used by Dubrovskii et al.

Table 5.1 Parameters used in fitting Dubrovskii's model to experimental data

Parameter	Starting Value (QD1)	Starting Value (QD2)
$a^{(s)}_{P,(1)}$	0.20	0.20
$a^{(s)}_{As,(1)}$	0.80	0.80
V_0	1.0	1.0
τ_{P}	−1.0	−1.0
τ_{As}	−2.0	−2.0
$t_{(1)}$	11.0	7.0
$t_{(2)}$	60.5	7.0
$t_{(3)}$	15.0	17.2

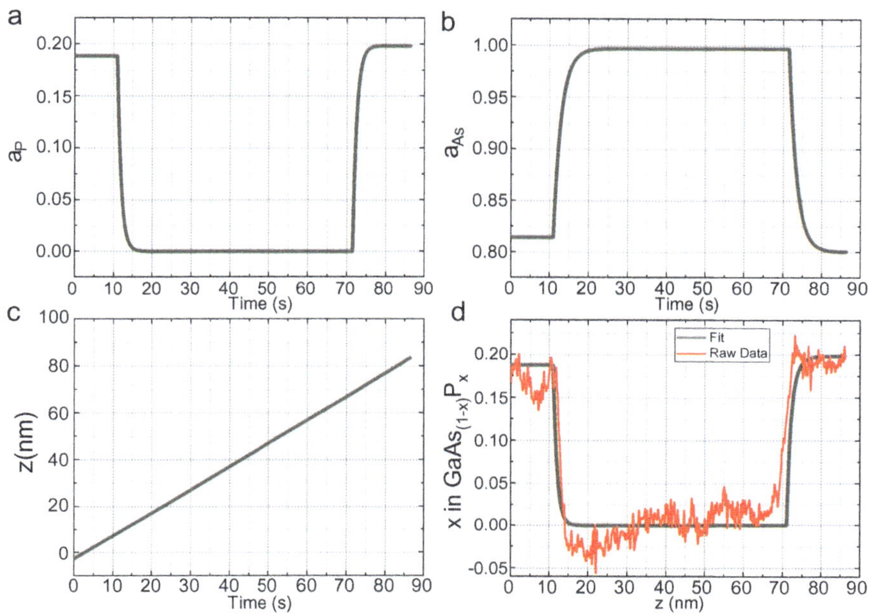

Fig. 5.6 **a, b** Material flux variables a_P and a_{As} used in the Dubrovskii model of NW growth plotted against time, produced from fitting the model to data. **c** NW height z plotted against time. **d** The resulting composition profile fit produced using the Dubrovskii model (black) and experimental data (red)

[4] and from some experimental values. Initial a_i values are estimated based on composition of experimental composition profiles. During QD growth (Sect. 2), a constraint used is when the P flux is turned off, the As flux is adjusted such that total group V flux is constant. From known approximate times of growth (Fig. 5.1a) and measuring QD features grown, V_0 is estimated to be $(1 \pm 0.1)\,\text{nms}^{-1}$.

An example of the fits produced for $a_P(t)$, $a_{As}(t)$, $z(t)$, and the resulting composition profile $x(z)$ with experimental data are shown in Fig. 5.6a–d respectively. This shows an example of one of the large QDs at the top of a NW (Fig. 5.1a) from the sample. The fit to experimental data successfully captures the general change in experimental parameter values. Values of τ from the fit are found to be $\tau_P = -1.0$ and $\tau_{As} = -2.0$. As can be observed in Fig. 5.6d this model provides a good fit to the experimental profile.

To evaluate how well the model fits the data and compare to other models, residuals and their distributions were extracted, where a residual mean ≈ 0 indicates a good fit [18]. An example of this analysis is shown in Fig. 5.7, containing the information for an approximately 10 nm long QD. Parameter starting values used to obtain this fit are given in column QD2 in Table 5.1. This fit using Dubrovskii's model presents the expected shape for this type of interface, with the model fit capturing the profile

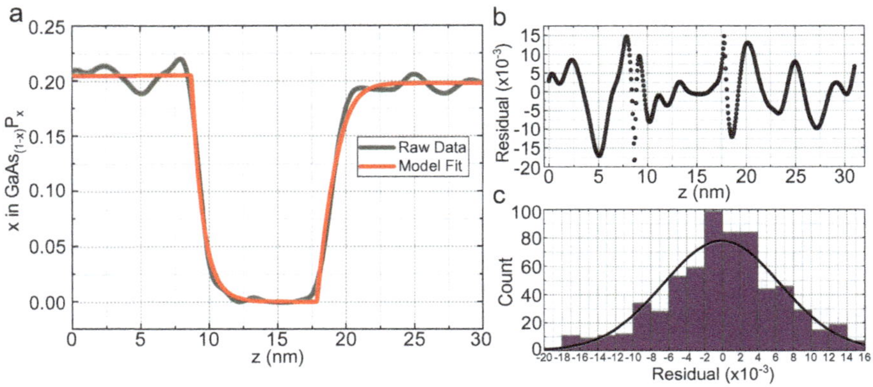

Fig. 5.7 Fitting the Dubrovskii kinetic model to a QD. **a** Model fit (red) and experimental data (black). **b** Residual of the fit produced compared to the experimental data. **c** Histogram to show the distribution of the residuals with a normal distribution fit (black curve)

shape well. Figure 5.7b, c correspond to the residuals from the fit to Dubrovskii's model and a normal distribution fit to a histogram of the residuals respectively. The residual values and distribution indicate a good fit to the data, with residual distribution centred at -5.5×10^{-5} and $\sigma = 0.0066$.

5.3.2 Comparing Models

To compare all the models being considered here, the data are limited to one interface at a time. This is a requirement for some of the models considered as they only deal with the interface itself. Dealing only with the interface is potentially a source of error since defining where the interface starts is required. This is difficult because of both the reservoir effect which fades the projected image of the interface and also the strain effects that influence ADF-STEM image intensity at interfaces. This is one aspect which makes the model proposed by Dubrovskii more representative of physical effects that influence the composition profile of heterostructures in NWs.

The left GaAsP/GaAs interface of the QD shown in Fig. 5.7 is used to compare the four models considered in this section. The experimental data is shown in Fig. 5.8a by the solid black line. The graph also contains four different fittings applied to the data: Dubrovskii model (purple dashed line), Priante model (red dashed line), Muraki model (green dashed line), and empirical Boltzmann model (blue dashed line).

For the Muraki fit, Eq. 1.33 was used to fit the profile,

$$x_n = x_0(1 - R^N)R^{n-N} \tag{5.4}$$

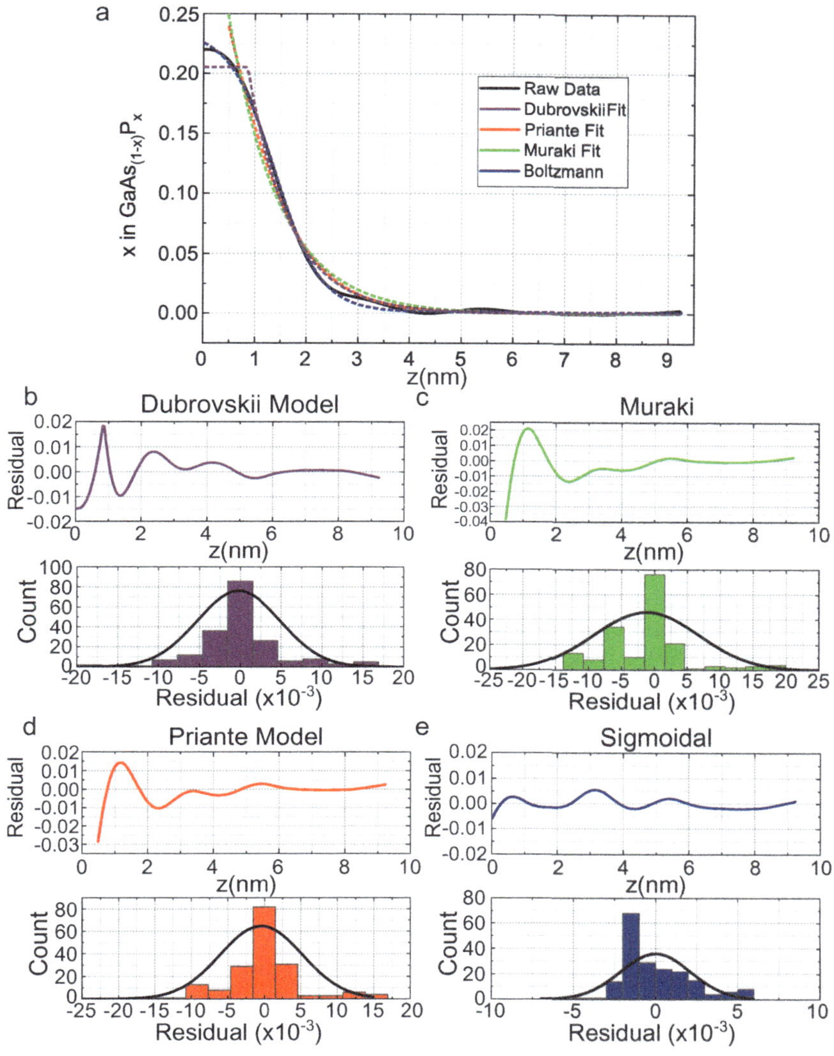

Fig. 5.8 **a** The left interface from Fig. 5.7 with 4 different fits labelled. **b–e** Residuals and their distributions for each of the labelled models. Normal distributions fit to the residuals are shown by the black curves

The fit to experimental data is represented by the green dashed line in Fig. 5.8a giving parameter values: $x_0 = 0.73$ $R = 0.722$ and $N = 1.35$.

Equation 1.28 from the Priante et al. model (Sect. 1.4.2) is used to fit composition against position along the NW (z), and is shown in Fig. 5.8a by a red dashed line. Parameters ϵ, y_0, and g were allowed to vary in the fitting process.

Table 5.2 Comparing the distributions of the residuals from interface models tested

Model	Centre	σ
Dubrovskii	-1.91×10^{-4}	0.0051
Priante	-4.22×10^{-4}	0.0056
Muraki	-12.0×10^{-4}	0.0079
Sigmoidal	-5.40×10^{-14}	0.0021

For the empirical Boltzmann sigmoidal fit, Eq. 1.36 was used to fit the data and is repeated here for convenience

$$y = \frac{A_1 - A_2}{1 + e^{(x-x_0)/dx}} + A_2 \tag{5.5}$$

The fit shown by the blue dashed line in Fig. 5.8a gives parameter values $A_1 = 0.236 \pm 0.001$, $A_2 = 0.00171 \pm 0.00002$, $x_0 = 4.24 \pm 0.02$, and $dx = 1.36 \pm 0.01$.

To compare the model fits against the data, the residuals and distributions from $Data - Fit$ are analysed. Normal distribution fits of the residual data were used to quantify residuals. Separate plots for each model are shown in Fig. 5.8b–e, and the normal distribution parameters are given in Table 5.2. The differences between the goodness of fit for the different models are small. Looking at the distributions of residuals, the Dubrovskii model shows the smallest residual mean and σ of the three physical models. The residuals of the Priante and Muraki model show very similar trends, with the Priante model a slightly better fit. Considering just this GaAsP/GaAs interface alone, the empirical sigmoidal model provides the best fit, with the smallest residuals distributed close to zero. It can be concluded that while the empirical sigmoidal model provides the best fit to the data, Dubrovskii's model provides the best fit of the physical models. The Muraki model was the weakest of the physical models, as expected from the Muraki model which is designed to describe segregation in VS growth and not VLS growth. Dubrovskii's model benefits from not being limited by the choice of where the interface starts.

5.4 Interface Width

In the previous section, it was inferred that while the physical models based on NW growth can more accurately describe composition profiles of heterostructures in NWs, the empirical sigmoidal fit provides a good fit for this type of interface and can be used to determine the interface width. In this section, the relation between axial heterostructure interface width and NW radius is examined. Values of interface width have been determined in two different ways:

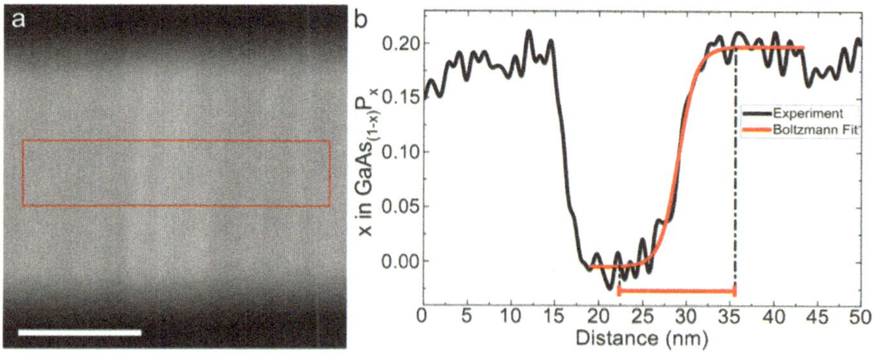

Fig. 5.9 **a** ADF-STEM image of a QD in a NW with the red box indicating where an intensity profile is taken and shown in (**b**). A Boltzmann sigmoidal fit is made to the right interface. The red line towards the bottom of the graph indicates the width of interface. Scale bar 20 nm

1. Measurement of the distance between points at 2 and 98% of the maximum composition values, i.e. the distance between 0.004 to 0.196 P composition (after background removal via linear interpolation and subtraction of ADF intensity across the QD).
2. Using the sigmoidal fit to determine the gradient in composition change using Eq. 1.37. A larger gradient corresponds to a narrower interface.

An example of using both methods is shown in Fig. 5.9. Figure 5.9a is an ADF image of a NW with the red box indicating the area used for the composition profile based on intensity (black line in Fig. 5.9b). In Fig. 5.9b, the red line corresponds to the Boltzmann fit for the P composition. Following the two different methods described above to determine interface width, method (1) in this case gives a width of 11.8(5) nm. Method (2) gives parameters $A_2 = 0.1977$, $A_1 = -0.0043$, $dx = 1.05$ nm and from Eq. 1.37 the gradient of the central straight section is $(0.048 \pm 0.002)\ \text{nm}^{-1}$.

The QD heterostructure inside the NWs have two different interfaces, interface type (a)—GaAsP/GaAs and interface type (b)—GaAs/GaAsP (i.e. the lower and upper interface with respect to growth direction). Compositional profiles were extracted using images similar to Fig. 5.9. The profiles were split into type (a) and (b) interfaces, which is necessary in order to fit the Priante and Muraki models to the experimental data. Additionally the data is divided into QDs with and without twinning, since twins indicate sub-optimal growth conditions leading to different growth behaviour. Sub-optimal growth conditions indicate geometrical parameters such as contact angle β vary during growth, and as was seen in Eq. 1.19, the relaxation time τ depends on geometrical parameters. A change in geometrical parameters will then change τ and influence interface sharpness (see Sect. 1.4.1). Separating the data into QDs with and without twinning allows for the analysis of interface sharpness to be split into sharpness influenced by radius and droplet geometry, and just radius respectively.

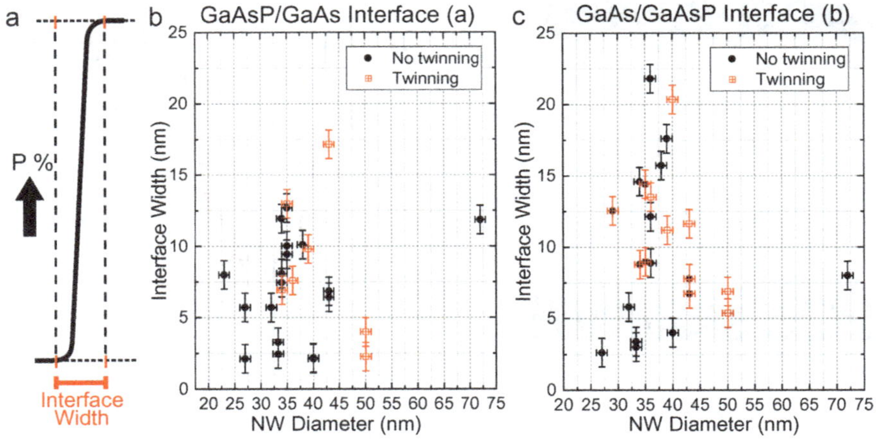

Fig. 5.10 **a** Schematic to show how width of interface (red line) is determined based on intensity values. **b** Plot of interface width from GaAsP/GaAs (a) interfaces against NW diameter with the data split into QDs with and without twinning. **c** A similar plot of interface width but this time for the GaAs/GaAsP (b) interface

The data collected using method (1) to determine interface width is shown in Fig. 5.10. Figure 5.10a is a schematic of the method used to measure the interface width. Figure 5.10b, c correspond to the interface widths for the type (a) and type (b) interfaces respectively. The black circles correspond to data from QDs without twinning and the red squares data with twinning. The trend expected is a sharper interface for thinner NWs as demonstrated experimentally [19] and theoretically [4, 5, 7]. While arguments could be made that these data show some indication of this, there is no conclusive evidence here that agrees with the expected trend for both interface types (a) and (b). The data from interface type (a) with no-twinning shows the best sign of the expected trend.

Figure 5.11 shows the same analysis using method (2) to determine interface sharpness, and is illustrated in Fig. 5.11a. The interface sharpness is determined based on the gradient of the section of the Boltzmann fit indicated by the red section of the line in Fig. 5.11a. Here the modulus of the gradient is presented so a larger number indicates a sharper interface for both interface types (a) and (b). Again, there is no obvious trend in the data. There are clear indications that the lower GaAsP/GaAs (a) interfaces in general are sharper than the upper GaAs/GaAsP (b) interfaces. This may be expected from the lower mobility of P atoms and the more efficient incorporation of As [20]. P has a shorter depletion time from its lower solubility in the catalyst and higher evaporation rate compared to As [14]. When the P flux is switched off, the As atoms will more readily be incorporated into the structure.

A reason for no obvious trends in the interface sharpness with NW radius could be the nature of the GaAsP/GaAs interfaces, which are already much sharper than other III-V/III-V interfaces [4, 14, 21, 22]. The GaP-GaAsP system is known to

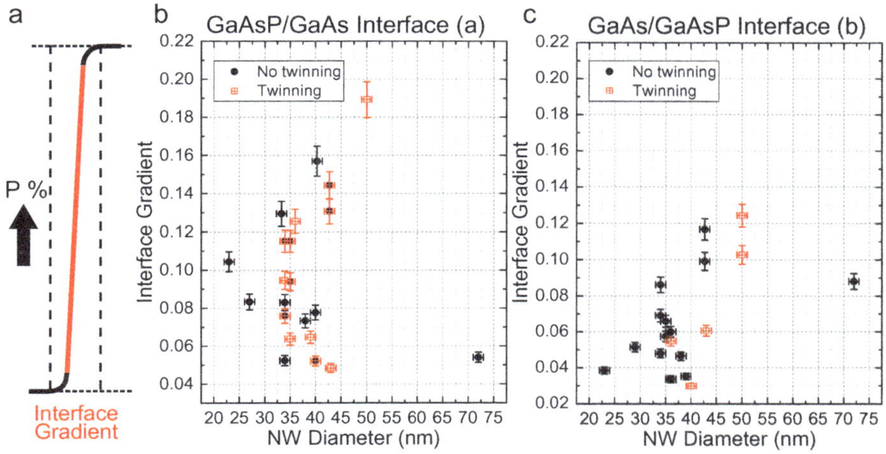

Fig. 5.11 **a** Schematic to show how sharpness of interface is determined based on the gradient of a sigmoidal fit to data. **b** Plot of modulus of interface gradient from GaAsP/GaAs (a) interfaces against NW diameter with the data split into QDs with and without twinning. **c** A similar plot of interface gradient for GaAs/GaAsP (b) interfaces

have sharp interfaces because of low group V concentration and solubility inside the catalyst [14, 23].

5.5 Size of Quantum Dots

While the main objective of the designed sample grown was to determine if there was a relation between NW radius and sharpness of heterostructure GaAsP/GaAs interfaces, extra information was extracted, i.e. the QDs size variation with NW radius for the same flux rate of material for a period of time. For a fixed flux of incoming material, NWs with larger radius would be expected to have thinner QDs. To analyse if there is any trend between the QDs size and the NW radius in the sample, the smallest and largest QD from the group of five QDs near the top of the NW were measured. QD size was determined by the distance between points at 50% of the maximum composition values at each interface.

Plots of QD size as a function of NW diameter are presented in Fig. 5.12 and are split into the largest dot (Fig. 5.12a) and smallest dot (Fig. 5.12b), with dot size referring to the designed varying QD sizes as was seen in Fig. 5.1. The data has been split into QDs without twinned planes (black circles) and those with twinning (red squares). They have been split in this way since twinning is a sign of changes in energetics of growth and so may be a sign of differing growth conditions for a particular NW. For both plots, there is no obvious trend visible, however it could be argued there is an expected trend of smaller QDs seen for thicker NWs.

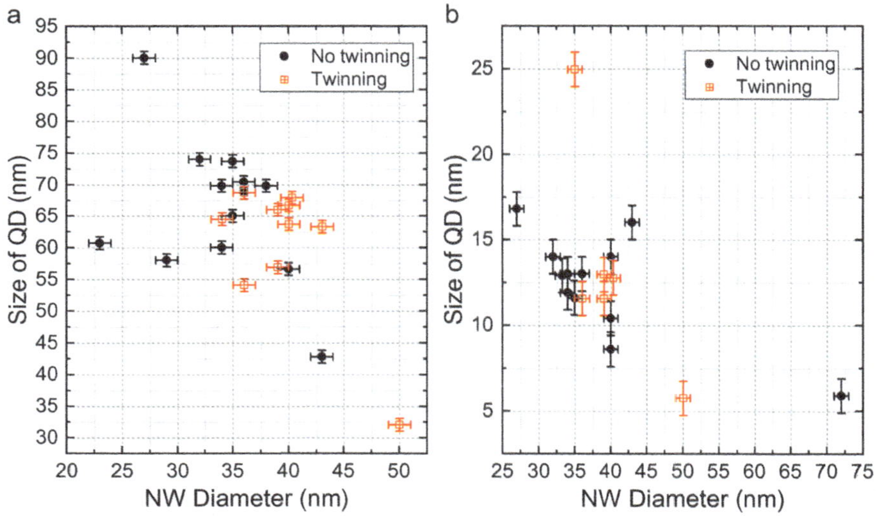

Fig. 5.12 Axial size of QDs for different diameters of NW. **a** Shows QD size for the top largest QD as seen in Fig. 5.1a and **b** the size of the smallest QD

There can be a few reasons for no obvious trend. Some effects that influence QD size are the same effects that cause length distribution in self-catalysed NW growths. Factors that influence length distribution of NWs include (i) nucleation delay of NWs growing from the substrate [24], (ii) kinetic fluctuations [25–28], (iii) diffusion-induced broadening [29], (iv) shadowing effects [30] and (v) secondary nucleation of Ga droplets [28]. The NW shape is also likely to be an influencing factor in length distribution, with NWs not perfectly hexagonal, as shown in Chap. 4. Any twins present in the structure also suggest changes in the growth process. It was previously reported that ZB and WZ sections were observed to grow at different rates in GaP NWs with GaAsP QDs [23], and this could contribute to variation in QD size. The length distributions have been reported for GaAs NWs [31] and length distributions have also been reported to differ depending on NW radius, with a smaller radius leading to a broader length distribution [29]. The length distribution changing for different radii together with the different contributing factors for length distribution can lead to a range of QD sizes for different radii, and this is seen in Fig. 5.12. To get a clearer picture of what is happening with QD size for different NW radii, a larger sample size is required in order to analyse size distributions for different radii.

5.6 Chapter Summary

This chapter has looked into how the NW radius influences interfaces of axial heterostructures in self-catalysed GaAsP NWs. To do this a structure was designed and grown with NWs of varying radius sizes with a pattern of systematic insertions of five GaAs QDs repeated five times along the GaAsP NW.

It was seen in these NW structures that strain can influence image intensity. Strain from the change in composition across QD boundary leads to a dip in image intensity close to the interface from dechannelling. This effect is amplified if the sample is tilted, and so when looking at interface profiles care must be taken whether the NW is tilted or not as this can unintentionally alter compositional analysis performed based on image intensity.

A selection of models were examined that model composition variation across a heterostructure interface, with two of these being models built around the NW system, namely Dubrovskii's kinetic model and Priante/Glas' thermodynamic model. Two additional models were considered; the Muraki segregation model commonly used in thin film systems, and an empirical sigmoidal model. The Dubrovskii model was seen to best model NW heterostructures.

An investigation was made into how sharpness of heterostructure interfaces changes with NW radius. There was no obvious relation between interface sharpness and NW radius, and is attributed to low group V concentration inside the catalyst [23]. A more obvious trend may be expected in a system such as GaAsSb-GaAs for example.

Finally, the size of two QDs was measured for different NW radii. While no clear conclusions could be made, there were signs of an expected trend of larger dots in thinner NWs. The observation of no obvious trend in QD size is attributed to length distribution effects that cause variation in NW length, in this case applied to QDs.

References

1. Kelly MJ, Palmer RE, Lauhon LJ, Gudiksen MS, Lieber CM (2004) Semiconductor nanowire heterostructures. Philos Trans R Soc Lond Ser A Math Phys Eng Sci 362:1247–1260
2. McIntyre P, Fontcuberta i Morral A (2020) Semiconductor nanowires: to grow or not to grow? Mater Today Nano 9:100058
3. Jia C, Lin Z, Huang Y, Duan X (2019) Nanowire electronics: from nanoscale to macroscale. Chem Rev 119:9074–9135
4. Dubrovskii VG, Sibirev NV (2016) Factors influencing the interfacial abruptness in axial III-V nanowire heterostructures. Cryst Growth Des 16:2019–2023
5. Priante G et al (2016) Sharpening the interfaces of axial heterostructures in self-catalyzed AlGaAs nanowires: experiment and theory. Nano Lett 16:1917–1924
6. Glas F (2017) Comparison of modeling strategies for the growth of heterostructures in III-V nanowires. Cryst Growth Des 17:4785–4794
7. Understanding the composition of ternary III-V nanowires and axial nanowire heterostructures in nucleation-limited regime. Mater Des 132:400–408 (2017)

8. Zhang Y et al (2016) Influence of droplet size on the growth of self-catalyzed ternary GaAsP nanowires. Nano Lett 16:1237–1243
9. Carapezzi S et al (2014) Bundling of GaAs nanowires: a case of adhesion-induced self-assembly of nanowires. ACS Nano 8:8932–8941
10. Williams D, Carter C (2009) Transmission electron microscopy a textbook for materials science. Springer
11. Brydson R (2011) Aberration-corrected analytical transmission electron microscopy. Wiley
12. Beyer A et al (2017) Influence of surface relaxation of strained layers on atomic resolution ADF imaging. Ultramicroscopy 181:8–16
13. Pantzas K et al (2012) Nanometer-scale, quantitative composition mappings of In-GaN layers from a combination of scanning transmission electron microscopy and energy dispersive x-ray spectroscopy. Nanotechnology 23:455707
14. Priante G, Patriarche G, Oehler F, Glas F, Harmand J-C (2015) Abrupt GaP/GaAs interfaces in self-catalyzed nanowires. Nano Lett 15:6036–6041
15. De Backer A, van den Bos K, Van den Broek W, Sijbers J, Van Aert S (2016) StatSTEM: an efficient approach for accurate and precise model-based quantification of atomic resolution electron microscopy images. Ultramicroscopy 171:104–116
16. Rosenauer A et al (2011) Composition mapping in InGaN by scanning transmission electron microscopy. Ultramicroscopy 111:1316–1327
17. Grillo V (2009) Quantitative evaluation of strain effects in STEM HAADF contrast. Microscopie 11:61–68
18. Hughes I, Hase T (2010) Measurements and their uncertainties. Oxford University Press
19. Li N, Tan T, Gösele U (2008) Transition region width of nanowire hetero- and pn-junctions grown using vapor-liquid-solid processes. Appl Phys A 90:591–596
20. Shu-Dong W et al (2005) Incorporation behaviour of arsenic and phosphorus in GaAsP/GaAs grown by solid source molecular beam epitaxy with a GaP decomposition source. Chin Phys Lett 22:960–962
21. Gudiksen MS, Lauhon LJ, Wang J, Smith DC, Lieber CM (2002) Growth of nanowire superlattice structures for nanoscale photonics and electronics. Nature 415:617–620
22. Borgström MT, Verheijen MA, Immink G, de Smet T, Bakkers EPAM (2006) Interface study on heterostructured GaP-GaAs nanowires. Nanotechnology 17:4010–4013
23. Jabeen F, Patriarche G, Glas F, Harmand J-C (2011) GaP/$GaAs_{1-x}P_x$ nanowires fabricated with modulated fluxes: a step towards the realization of superlattices in a single nanowire. J Cryst Growth 323. Proceedings of the 16th international conference on molecular beam epitaxy (ICMBE), pp 293–296
24. Dubrovskii VG, Barcus J, Kim W, Vukajlovic-Plestina J, i Morral AF (2019) Does desorption affect the length distributions of nanowires? Nanotechnology 30:475604
25. Dubrovskii VG (2009) Fluctuation-induced spreading of size distribution in condensation kinetics. J Chem Phys 131:164514
26. Dubrovskii VG (2016) Kinetic narrowing of size distribution. Phys Rev B 93:174203
27. Glas F, Dubrovskii VG (2017) Self-narrowing of size distributions of nanostructures by nucleation antibunching. Phys Rev Mater 1:036003
28. Tauchnitz T et al (2018) A simple route to synchronized nucleation of self-catalyzed GaAs nanowires on silicon for sub-Poissonian length distributions. Nanotechnology 29:504004
29. Dubrovskii VG et al (2016) Length distributions of nanowires growing by surface diffusion. Cryst Growth Des 16:2167–2172
30. Koivusalo ES, Hakkarainen TV, Guina MD, Dubrovskii VG (2017) Sub-poissonian narrowing of length distributions realized in Ga-catalyzed GaAs nanowires. Nano Lett 17:5350–5355
31. Vukajlovic-Plestina J et al (2017) Engineering the size distributions of ordered GaAs nanowires on silicon. Nano Lett 17:4101–4108

Chapter 6
Conclusions and Future Work

This thesis has looked extensively at planar defects that can be found in self-catalysed III-V nanowire (NW)s when growth proceeds at sub-optimal conditions, particularly during catalyst droplet consumption. Defects have been identified through the use of atomic resolution electron microscopy enabled by probe corrected scanning transmission electron microscope (STEM). A large variety of defect types have been observed, including familiar partial dislocations with unfamiliar core structures. Unfamiliar core structures are a result of multiple twinning and inclined twins that interact with defects. The presence of defect structures in NWs can diminish optoelectronic performance and so identifying the exact type of defects can help to guide efforts made in reducing their prevalence.

Chapter 3 presented a catalogue of defects that have been observed in self-catalysed NWs, with analysis of Burgers vectors performed to categorise each defect structure, including unfamiliar core structures of familiar defects where the structure is multiply twinned. Defects observed include intrinsic and extrinsic 30° Shockley partial dislocations, Lomer-Cottrell lock (LCL)s, and the most commonly observed defect structure was the $\Sigma 3\{112\}$ twin boundary, referred to as the 3monolayer (ML) defect that has a zero Burgers vector. The question of how the defects may be forming in growth was addressed using in-situ heating experiments inside a transmission electron microscope (TEM), and showed it is possible for two islands of new material to nucleate during growth, and provided one of these is twinned then defects may form. Evidence for the detrimental effect on optoelectronic properties was seen using cathodoluminescence (CL) analysis, where no emission was seen in the defective tip region of NWs. By taking cross-section slices from the tip region of defective NWs, there is evidence that defects can extend through the entire thickness of the NW, they show signs of kinks, or they can interact with other defects in the thickness of the NW.

A more thorough study of some of the defects observed with unfamiliar core structures could be performed for example using density functional theory (DFT) to see how the core structure affects band structure. Model defects with these core structures could also be produced and relaxed to see how bonding may be different for

J. A. Gott, *Defects in Self-Catalysed III-V Nanowires*, Springer Theses,
https://doi.org/10.1007/978-3-030-94062-1_6

these core structures. While evidence for how the defects observed may be forming during NW growth from in-situ studies, this is one area where more work would help to build up a better picture of what is happening during NW growth when defects form. Currently in the literature there are studies on good growth conditions, and so one area that could be investigated is in-situ studies of NW growth with sup-optimal growth conditions. This could help to determine the limits of optimal growth conditions. To my knowledge there have been no reports of in-situ studies of the droplet consumption stage of growth, and while this would be a difficult and time consuming experiment, such an experiment would help to the further understanding of NW growth. In-situ studies could also be performed on vapour-solid (VS) growth of NWs, since efforts so far have been in vapour-liquid-solid (VLS) NW growths. The defects presented in Chap. 3 were all from the same GaAsP material system, it may be interesting to check if similar observations of types of defects and their prevalence is true in other binary or ternary III-V systems. Further work could also be done in the analysis of cross-sections of defective NWs. While some of the explanations for contrast features seen in the NW cross-sections have some supporting evidence, more work should be done to confirm where contrast in the cross-section images is coming from. Cross-sections could for example be tilted to a larger range of tilt angles, or tomography analysis could be performed, which could reveal a more complete picture of where inclined twins are located and any defects present that can contribute to the contrast features seen in the NW cross-sections.

Chapter 4 looked into how defects observed in self-catalysed NWs move, or in some cases remain stable, when the temperature of the NW is elevated. The motion of the most commonly observed 3ML type defect was the main focus of this chapter. The driving force behind motion of 3ML defects comes from reduction of Gibbs free energy of the system by reducing the area of twinned interfaces. Exact location and shape of the NW influence the strength of forces acting on the defects. Behaviour of 3ML defects observed included (i) complete removal from the system, (ii) motion that is stopped by interaction with another defect locking it in place, and (iii) no motion from a stable configuration. The expected trend of higher defect velocities at higher temperatures was observed, although a limited number of high velocity examples were observed since defects were removed from the system before high velocities could be measured. It was seen that up to 70% of the defects present in the studied system could be removed from the NW system by annealing, and shows that annealing NWs after droplet consumption during growth can be an easy way to remove defects present in the NW that diminish optoelectronic properties, and improve material quality.

In the images presented in Chap. 4 where defects are in the process of moving, the blurred appearance has been attributed to kinks. The blurred appearance can vary in size, and this is assumed to be from multiple kinks at different positions along the defect line. The exact reason for the appearance could be examined further by performing a series of simulations that test how the appearance of defects change based on the position of kink along the defect line. This could be taken further by, for example, performing molecular dynamics simulations of defect motion and produce STEM images based on these simulations to see how well they match experimental images. Temperatures used in the experiments of Chap. 4 were relatively high compared to the growth temperature so that more obvious motion should occur in a short amount of time. A similar in-situ annealing experiment using lower temperatures could be done to check for the lowest temperature required for a reasonable response from defects to provide better guidance for post growth anneal conditions for defect removal. It was seen in Chap. 3 from CL mapping that there is no emission observed from the tip region of NWs and is attributed to the presence of defects causing non-radiative recombination. To test if annealing NWs can remove defects and recover emission, an investigation could be made into taking CL maps before and after annealing and comparing them.

Chapter 5 examined heterostructure interfaces in self-catalysed GaAsP NWs with axial GaAs quantum dots (QDs) from NWs of varying radius. Different models that describe the compositional variation across a heterostructure were considered, and included two models recently developed specifically for NWs, namely Dubrovskii's model and the model by G. Priante & F. Glas. Other models considered included the Muraki model and an empirical sigmoidal model. Fitting the models to experimental data showed Dubrovskii's model to be the best fit of the physical models tested. By using sigmoidal fits to experimental interfaces, the interface sharpness was measured for different NW radii. There was no obvious trend observed in the data, which is attributed to the material system grown tending to possess relatively sharp interfaces. The size of QDs was also measured as a function of NW radius. There was no obvious observable trend and this is attributed to NW growth leading to a Poisssonian distribution of QD sizes similar to length distributions observed in NWs.

The variation in interface thickness and QD size could be further tested by growing a similar sample of NWs with different radii with a passivating shell so that optical emission could be tested by using CL. Variation in interface sharpness and QD size would result in spectral broadening of emission, and so CL spectra could be acquired for NW of different radii to look for this. A similar sample of NWs with varying radius could be grown for a different III-V system to test the interface sharpness for different radii. GaAsSb for example would be expected to show a more obvious trend. To test if the size of QD follows a Poisssonian distribution a larger sample of NW radius and QD size measurements would be required to enable such analysis.

The manufacturer's authorised representative in the EU is Springer Nature Customer Service Centre GmbH, Europaplatz 3, 69115 Heidelberg, Germany. If you have any concerns regarding our products, please contact ProductSafety@springernature.com

Printed and bound by CPI Group (UK) Ltd, Croydon, CR0 4YY
28/09/2026
02224089-0018